Plants fro

APR 23 2019

Plants from the Past

Leonard W. Blake
and
Hugh C. Cutler

With an Introduction by
Gayle J. Fritz and Patty Jo Watson

THE UNIVERSITY OF ALABAMA PRESS
Tuscaloosa and London

Copyright © 2001
The University of Alabama Press
Tuscaloosa, Alabama 35487-0380
All rights reserved
Manufactured in the United States of America

2 4 6 8 9 7 5 3 1 / 00 02 04 06 08 07 06 05 03 01

Typeface: Minion

∞

The paper on which this book is printed meets the minimum requirements of American National Standard for Information Science–Permanence of Paper for Printed Library Materials, ANSI Z39.48-1984.

Library of Congress Cataloging-in-Publication Data

Blake, Leonard W.
    Plants from the past / by Leonard W. Blake and Hugh C. Cutler ; with an introduction by Gayle J. Fritz and Patty Jo Watson.
      p. cm.
    Includes bibliographical references (p.   ) and index.
    ISBN 0-8173-1087-8 (alk. paper)
    1. Plant remains (Archaeology)   2. Excavations (Archaeology)—United States.   3. United States—Antiquities.   4. Paleoethnobotany.   5. Indians of North America—Antiquities.   6. Corn—United States—History.   7. Plants, Cultivated—United States—History.   I. Cutler, Hugh C., 1912–   II. Title.
CC79.5.P5 B58 2001
930.1—dc21   2001001288

British Library Cataloguing-in-Publication Data available

Contents

List of Illustrations   vii

Acknowledgments   ix

Introduction   xi

Map of Site Locations   xv

1. North American Indian Corn   1
   HUGH C. CUTLER AND LEONARD W. BLAKE, 1976

2. Cultivated Plants from Picuris   19
   HUGH C. CUTLER, 1966

3. Corn in the Province of Aminoya   37
   LEONARD W. BLAKE, 1974

4. Corn from Three North Carolina Sites, 31Gs55, 56, and 30   40
   LEONARD W. BLAKE, 1987

5. Cultivated Plant Remains from Historic Missouri and Osage Indian Sites   46
   LEONARD W. BLAKE, 1986

6. Corn for the Voyageurs   54
   LEONARD W. BLAKE, 1994

7. Corn from Fort Michilimackinac, A.D. 1770–1780   59
   LEONARD W. BLAKE AND HUGH C. CUTLER, 1968

8. Corn from the Waterman Site (11R122), Illinois   66
   LEONARD W. BLAKE, 1972 (REVISED 1997)

9. Plant Remains from the Rhoads Site (11Lo8), Illinois   72
   LEONARD W. BLAKE AND HUGH C. CUTLER, 1974

10. Plants from Archaeological Sites East of the Rockies   93
    HUGH C. CUTLER AND LEONARD W. BLAKE, 1976

11. Published Works of Cutler and Blake   148

Works Cited   157

Index of Latin Names for Plant Taxa   165

Index of Corn Races and Varieties   169

General Index   171

Illustrations

Figures

2.1.	Corn Cobs from the Taos Phase, A.D. 1150–1225	21
2.2.	Corn Cobs from the Santa Fe Phase, A.D. 1225–1300	29
2.3.	Corn Cobs from the Vadito Phase, A.D. 1375–1490	30
2.4.	Corn Cobs from the Trampas Phase, A.D. 1600–1696, Area VI	31
2.5.	Corn Cobs from the Trampas Phase, A.D. 1600–1696, Area II	32
2.6.	Corn from Picuris in 1953 and 1963	33
2.7.	Maize from Two Reservations, 1953	34
4.1.	Corn from 31Gs55, North Carolina	41
4.2.	Corn from 31Gs56, North Carolina	43
4.3.	Corn from 31Gs30, North Carolina	44
5.1.	Measurable Squash Seeds from Historic Missouri and Osage Sites	51

Tables

2.1.	Percent of Cobs of Each Row Number Found in Selected Southwestern Sites	26
2.2.	Thirty-nine Ears of Corn from Modern Picuris (1953 and 1963)	28
4.1.	Comparative Samples of Corn from North Carolina, Virginia, and Georgia	45
5.1.	Comparison of Corn Cobs from Historic Missouri and Osage Sites and from a Historic Kickapoo Site	47
5.2.	Beans (*Phaseolus vulgaris*) from Historic Missouri and Osage Sites and Three Other Historic Sites	49
7.1.	Comparative Data on Northern Flint Corn from Selected Archaeological Sites	62
8.1.	Corn Cobs from Different Locations on the Waterman Site	67
8.2.	Comparisons of Corn Cobs from the Waterman Site (11R122) with Those from Other Indian Sites of about the Same Time Period	68
9.1.	Comparisons of Corn Cobs from the Rhoads Site (11Lo8) with Selected Midwestern Sites	76

9.2.	Beans from the Rhoads Site (11Lo8) Compared with Those from Other Sites 79
9.3.	Beans (*Phaseolus vulgaris*) from the Rhoads Site (Carbonized) 80
9.4.	Measurements of Squash (*C. pepo*) Seeds from the Rhoads Site (11Lo8) 81
9.5.	Watermelon (*Citrullus lanatus*) Seeds from the Rhoads Site (11Lo8) Compared with Those from Other Historic and Protohistoric Sites 83
9.6.	Comparisons of Plant Remains from Subclasses of Class 1 Pits at the Rhoads Site 87
9.7.	Summary of Cultivated Plants from Class 1 Pits on the Western Half of the Rhoads Site (11Lo8) 89
9.8.	Approximate Harvest Times of Cultivated and Wild Plants 91

Acknowledgments

I would like to heartily concur with the acknowledgments of Gayle Fritz and Patty Jo Watson in their introduction and, in addition, thank them for their contribution, without which there would have been no report. I also wish to express my gratitude to the late Hugh Cutler, a valued friend, who opened up a whole new world of ideas and interests. Thanks are also due to many helpful friends, and to those who have provided plant remains for study. These include Bob Bray, James Brown, Margaret Brown, Carl Chapman, Janet Levy, Alan May, M. D. Thurman, and the Illinois State Museum.

Leonard Blake

Introduction

Gayle J. Fritz and Patty Jo Watson

Hugh Cutler and Leonard Blake worked together on archaeological plant remains for a quarter century at the Missouri Botanical Garden. During that time they published dozens of articles and reports and provided many more unpublished responses to various archaeologists. After Cutler's retirement in 1977, Blake continued for two more decades. Some of their significant works, however, were not widely circulated, or for various reasons beyond their control did not make it into print. Those works are included in this volume, as is a complete bibliography of their archaeobotanical publications, and the full text of "Plants from Archaeological Sites East of the Rockies," their best-known compilation and one much sought after. This inventory was originally a mimeographed summary of species found at various sites whose excavators sent material to Cutler for identification. The mimeographed version was produced in 1973 in 200 copies (collated, stapled, and informally bound by Hugh, his wife Marian, and Leonard). They had mailed out all but a half dozen copies when, in 1976, the Missouri Archaeological Society offered to publish a microfiche edition. That version has long been virtually inaccessible. The full text of the 1976 update of "Plants from Archaeological Sites East of the Rockies" is available here in print (Chapter 10) for the first time.

Other works in this volume include a broad overview of corn grown by Native North Americans (Chapter 1), a playful calculation of field sizes and corn yields based on observations by early Spanish explorers in the Southeast (Chapter 3), a discussion of the sources of processed corn carried in canoes by early European traders in the upper Midwest and Canada (Chapter 6), and reports of archaeological plant remains from sites in New Mexico (Chapter 2), North Carolina (Chapter 4), Missouri (Chapter 5), Michigan (Chapter 7), and Illinois (Chapters 8 and 9). Blake has written introductory paragraphs to each chapter, explaining the circumstances under which the work was conducted and, if relevant, commenting on subsequent revisions in archaeological or botanical perspectives.

Most of these papers were written some time ago, but they have far more than historical value. The overview of North American Indian corn defines basic terms, clarifies racial classifications, and synthesizes archaeological, historical, and ethnographic information. Readers unfamiliar with Cutler's and Blake's breadth of knowledge will be impressed by their attention to social and ritual aspects of the use of corn and their excellent discussions of corn growing, storing, processing, and cooking. "Corn for the Voyageurs," written by Blake in 1994, presents unique and sophisticated insights into the logistics of provisioning early French trading expeditions. Several of the archaeobotanical reports contain valuable comparative data, not available elsewhere, from additional sites. Table 2.1, for example, in the chapter on cultivated plants from Picuris, gives a summary of row numbers of cobs from thirty-four components at sites across the Southwest, and Table 2.2 provides comparable data for cobs collected from modern Picuris in 1953 and 1963. Table 5.2, in the chapter on plants from Historic Missouri and Osage sites, summarizes measurements of beans from nine historic period sites in Missouri and Illinois. Corn from the three North Carolina sites presented in Chapter 4 is compared metrically with assemblages from Virginia and Georgia, strengthening Blake's suggestion that small cobs from 31Gs30 represent the early season corn mentioned in historic accounts. These types of information do not lose relevance with time. Seven chapters in this volume deal with assemblages or issues postdating European contact. This comes at a time of increasing attention by archaeologists and historians to early Indian-White interactions, making the work especially significant today.

Inevitably, some statements written as long as thirty-five years ago no longer reflect current thinking. Blake points out many of these in his introductory comments. People interested in recent evidence pertaining to the antiquity of cultigens in any given region or the evolutionary pathways for domestication of corn, beans, and squash need to consult current literature. Readers of this volume should be alert to temporal shifts in the views of Blake and Cutler between the 1960s and 1990s. For example, Eastern Eight Row corn was seen originally as having "dominated most of the region east of the Mississippi" by A.D. 1200 (Chapter 1). Later, however, it became evident that the dominance of Eastern Eight Row occurred primarily in the northern United States and Atlantic seaboard, whereas populations with higher average row numbers were common across much of the Midsouth and lower reaches of the Mississippi Valley even into post-Columbian times (Chapters 4 and 8).

Hugh C. Cutler earned a master of arts degree in botany at the University of Wisconsin (Madison) in 1936 and a doctorate (also in botany) from Washington University-St. Louis in 1939. In 1953 he became Curator of Useful Plants at the Missouri Botanical Garden in St. Louis, where he was employed until retirement in 1977, often acting in various administrative capacities as well as teaching ethnobotany to Washington University students. Cutler was instrumental in developing the technique of flotation to concentrate and recover charred botanical remains from archaeological deposits. Cutler was also for many decades a vigorous pioneer in the nascent subdiscipline of archaeobotany, especially the study of archaeologically derived corn, beans, and squash. He devised a technique for measuring cupule angles to enable estimates of row numbers from ancient cob fragments, and carried out a great deal of research on modern maize varieties in the U.S. Southwest, in Mexico, and in South America.

Leonard Blake earned a bachelor of arts degree from Williams College in 1927. For many years he was an investment securities analyst for the St. Louis Union Trust Company, a position from which he retired in 1965 at the age of 62. Beginning in the 1930s, he became an active avocational archaeologist. During the 1950s, he served as president of a local St. Louis avocational society, which—upon Hugh Cutler's invitation—held monthly meetings in the Museum Building of the Missouri Botanical Garden for a number of years. Several of these people, including Blake, aided Cutler with his archaeobotanical research. Eventually the group dwindled to three, then two, then only Blake was left. After his retirement, he established a regular two-day-a-week routine of volunteering in Cutler's lab at the Botanical Garden, a system he transferred to the archaeology lab at Washington University after Cutler retired in 1977 and moved to California. Blake maintains his Monday and Friday lab schedule to this day and is a vital resource for all archaeological students and faculty. He is 98 years old now, but his mental acuity is undiminished. He remembers with great clarity all the archaeological research in which he has participated, and he is a font of information on prehistoric and early historic archaeology of the greater St. Louis region.

Leonard Blake has received many honors during his long career as an avocational archaeologist. In 1982 he was presented with the first Distinguished Amateur Award by the Illinois Archaeological Survey; in 1983, the Missouri Archaeological Society gave him its Distinguished Service Award (including Life Membership); in 1985 he received an honorary doctor of science degree from Washington University; in 1986 Eric Voigt and Deborah Pearsall presented to him a festschrift edition of *The Missouri Archaeologist* entitled *New World Paleoethnobotany: Collected Papers in Honor of Leonard W. Blake*; in 1987, the Society for American Archaeology presented him with its Crabtree Award in recognition of his outstanding achievements as an avocational archaeologist and archaeobotanist; in 1994 the Missouri Association of Professional Archaeologists gave him the Chapmans' Award; and in 1999, he was presented with a Lifetime Volunteer Award by the Illinois Association for the Advancement of Archaeology.

Leonard Blake prepared this manuscript for publication because of our persistent encouragement (he would probably use a different term), but with only minimal editorial assistance from us. He is too humble to believe, but we know, that many students, professional and avocational archaeologists, and other people interested in past human-plant relationships will find value in these previously inaccessible works.

Several students at Washington University in St. Louis performed word processing and graph-making tasks that contributed greatly to the production of this book. Kimberly Schaefer skillfully transformed a diverse set of computer files and uncomputerized charts into stylistically consistent and publication-ready form. Gina Powell and Lucretia Kelly scanned or retyped several of the chapters. We are grateful to Judith Knight for her support and patience and to two anonymous University of Alabama Press reviewers for extremely helpful suggestions. Most of all, we express appreciation to Leonard Blake for his many contributions to ethnobotany and archaeology, and for years of sharing treasured knowledge and skills with us, our colleagues, and our students.

Map of Site Locations

Plants from the Past

1 North American Indian Corn

Hugh C. Cutler and Leonard W. Blake
Missouri Botanical Garden

(Originally written for *Handbook of North American Indians,* Environment, Origins, and Population volume, 1976)

Leonard Blake's Comments, 2000

In 1976, Dr. Cutler was asked to write a paper on the subject of North American Indian corn for inclusion in the volume on environment, origins, and population for the *Handbook of North American Indians,* which was projected by the Smithsonian Institution of Washington, D.C.

The paper was sent to Dr. Frederick S. Hulse, who was then editor for the projected volume. A number of years later, Dr. Cutler was notified that a publication date was approaching. An update was requested, in view of changes that had taken place. Dr. Cutler declined to do this, for in retirement, he had not attempted to keep up on recent developments. I have learned that a new author and a new editor have been chosen, but the projected volume for the *Handbook* has still not been published.

Hugh Cutler had an extensive knowledge of the corn of South America and, especially, that of Mexico and the American Southwest. This paper was written over twenty years ago and comments on Hopewell and similar Middle Woodland corn have proved to be valueless in the light of publication of corrected carbon 14 dates on corn obtained by the accelerator (AMS) dating technique, which first came into use in 1984. Before 1984, a certain minimum sample size was required for carbon 14 dating. Small samples of corn were usually dated by association, that is, by dates on material that was associated with them in the same feature. Accelerator dating, which permits the use of very small samples, showed that objects in the same feature were not always of the same age. Information on Hopewell and other Middle Woodland material has been obtained from three sites. One was the Jasper Newman site in Illinois, dated by association at 80 ± 140 B.C., which had an accelerator carbon 14 date of 450 ± 500 B.P. (Conard et al. 1984). The other two sites were in Ohio. The McGraw site had a date on bones of A.D. 230 ± 80, and the Daines II Adena mound was dated at 280 ± 140 B.C. Accelerator dates on corn from each of these sites came to less than 400 years B.P.

Some of the dates of corn's origins have also proved not as early as formerly thought. The earliest known corn from the Tehuacán Valley in Mexico, which was supposed to date at about 5000 B.C. (Mangelsdorf 1974:167), is now dated at no older than 3600 B.C. (Long et al. 1989). The earliest corn from Bat Cave, originally thought to date before 2000 B.C., was reassessed at 1200–1500 B.C. (Wills 1995).

In the paper, Cutler wrote the following statement: "Corn probably originated in west-central Mexico, a region where the largest populations of teosinte (*Zea mexicana* or *Euchlaena mexicana*) and many of the species of another closely related grass, *Tripsacum,* grow." Some may consider this to be a reasonable opinion, but the fact remains that the corn from the Tehuacán Valley is still, at this time, the earliest documented by archaeological excavations. Nowhere in this report is there any mention of Beadle's *The Mystery of Maize* (Beadle 1972), which, at the time, changed the viewpoint of many, but not all, corn experts on the origin of corn. The reason is that Cutler, like Mangelsdorf and some others, did not accept Beadle's conclusions.

Cutler wrote that a form of the Pima-Papago race of corn "spread eastward through Oklahoma and Arkansas and up the Mississippi River to reach northern Illinois by about A.D. 100, and southern Georgia perhaps even earlier." This is a very broad and perhaps not an accurate statement. In the accompanying "Plants from Archaeological Sites East of the Rockies," no northern Illinois site is shown with a date of A.D. 100. Some corn similar to Pima-Papago, which was called Tropical Flint (Anderson and Cutler 1942), has been found in Oklahoma, in Arkansas, at the Cahokia site, and at Mississippian sites on the upper Illinois River. Because of a lack of numerous, reliable early dates, just how and when it got there is not precisely known.

Maize, or Indian corn (*Zea mays*), has long been the most valuable plant in North American agriculture. It is the dominant grain of the region and is usually called *corn* in English-speaking areas of North America. *Corn* is the name applied in England and (as *Korn*) Germany to wheat, which is the most common grain there, and the name is used, in various other places, for oats, rye, barley, and even lentils.

Corn probably originated in west-central Mexico, a region where the largest populations of teosinte (*Zea mexicana* or *Euchlaena mexicana*) and many of the species of another closely related grass, *Tripsacum,* grow. Western Mexico is also the center for many wild species of squash and beans (*Cucurbita* spp. and *Phaseolus* spp.). Only a few of the many variants of Mexican corn traveled northward, their number reduced by barriers of arid lands and great distances between agricultural areas. Some diversity was introduced into the trickle of northbound corn by hybridization with teosinte, which grows wild and is tolerated as a weed or even planted with corn in some parts of northwestern Mexico. There may have been some mixing with *Tripsacum,* for *Tripsacum* occasionally exhibits characteristics similar to those of corn of the region and these grasses can cross (Wilkes 1971; de Wet and Harlan 1971). Usually in Mexico when teosinte is found, *Tripsacum* is growing nearby (Wilkes 1967).

The main route used by Indians traveling northward passes through Mexico's western mountains, into Arizona and New Mexico, and then to the east. Carl Sauer long ago (1941) mapped this route and pointed out the great diversity in cultivated plants and their relatives in the nearby mountains, and the propensity of botanists and other collectors to acquire specimens in large city markets and thus miss the diverse strains kept by small farmers, especially those still isolated in the northern part of the high basin country. Collections from such small farms of northwestern Mexico include primitive pop and flint ears and mixtures of these types with teosinte (Pennington 1969). Some of the ears resemble the oldest corn known from archaeological sites in North America (Anderson 1944, 1946; Wellhausen et al. 1952).

Compared to modern corn of the same region, corn from archaeological sites in northern Mexico is relatively uniform, from Durango to Tamaulipas (Mangelsdorf and Lister 1956; Brooks et al. 1962; Mangelsdorf, MacNeish, and Galinat 1967a). Seeds of teosinte, *Tripsacum*, and corn that resulted from hybridization with teosinte have been found in some of these sites. A very few specimens of the pyramidal dents common in central Mexico are found in later periods at a few sites.

The oldest corn in northern Mexico is from a southwestern Tamaulipas site believed to date from about 2350 to 1850 B.C. (Mangelsdorf, MacNeish, and Galinat 1967a). The oldest corn known, from caves near Tehuacán, Mexico, is considered to date from about 5000 B.C. (Mangelsdorf, Dick, and Comara-Hernandez 1967; Mangelsdorf 1974). Remains of bottle gourd and a cultivated squash were found in much older levels of nearby sites. In the United States the oldest archaeological corn comes from Bat Cave, a site just east of the continental divide in west-central New Mexico. This has been dated at about 2300 B.C. or several centuries later (Mangelsdorf, Dick, and Comara-Hernandez 1967) after consideration of carbon 14 dates of charcoal from nearby areas and comparisons of the corn with cobs from Tamaulipas sites. Slightly younger corn has been found at Tularosa Cave (Cutler 1952), which is only a few miles west of Bat Cave and across the continental divide. Both of these early corn-growing sites are on the edge of the ponderosa pines in country similar to that along the most traveled way from Mexico.

The Corn Plant

The structure of our corn plant follows the general pattern for lawn grasses, wheat, sugar cane, bamboo, and other grasses. The basic structure consists of a leaf with an associated bud and a section of stem. These units are arranged on alternate sides of an axis. The pattern extends to all parts of the grass plant, so the tassel and ear of corn, and the head, or ear, of wheat are modifications of a pattern readily seen in the stalk. Parts of an ear or tassel may vary in size and shape, there may be many more parts, and these may be condensed and tightly packed, but the basic pattern of a chain of units on alternate sides of an axis is maintained (Cutler and Cutler 1948).

The ear is borne on the end of a much-compressed branch of the corn plant. Each of the many husks is a modified leaf protecting the ear. In most other grasses the individual grains are concealed and protected by glumes and other parts of the spikelet. Kernels of primitive corn probably were covered by glumes. This protection is unnecessary when the entire ear is protected by husks, however, and is undesirable from the point of view of the harvester because it makes removal of the edible part difficult.

Seed Selection

Removing the glumes that covered the grains of wheat, rye, or barley always broke up the head. A wheat farmer usually sees his harvest as a mass of small and usually dull-colored grains and cannot select the ones that came from the largest heads. When cornhusks are removed, the farmer sees many large and bright kernels. There may be fewer than 200 on small Eastern eight-rowed corn or more than 800 on a dent corn from Jemez Pueblo. Each kernel on an ear is the progeny of the female parent and any one of many possible pollen parents. The visual impact of size, shape, texture, and especially color is magnified by the crowding of many kernels.

In the Old World, heads of grain were used for ceremonial purposes and ornament, but the character of the grains within the head was not apparent. In contrast, in North America, corn ears for ceremonial purposes were carefully selected largely because of the appearance of the kernels. At the birth of a child, for healing, or for harvest ceremonies, selected ears were used. These ears were carefully kept in the safest place, and the kernels planted, often in separate plots. There are reports of selection of special or "perfect" ears among the Rio Grande Pueblos, the Plains Indians, and farmers of the Eastern Woodlands. Because there are so many kernels on each ear (about 500 on a good ear of Hopi white or blue corn) such ceremonial ears would contribute a large proportion of the seed, besides establishing standards for the selection of other seed for planting.

A farmer who plants white flint kernels and harvests ears of mixed color and texture is likely to be curious. The white flint could also carry other, and invisible, recessive characters. When young ears grow, each ovary produces a silk that may be pollinated by a wind-blown pollen grain from the farmer's own plants or from a neighbor's.

The outer layer of a kernel, called the pericarp, is a tissue of the ear-bearing plant. Pericarp may be colorless, yellow, brown, various shades of red, or striped with red. Directly within this is the aleurone, usually a single layer of cells, which receives two sets of chromosomes from the female parent and one set from the pollen, or male, parent. Aleurone may be colorless, yellow, brown, lilac, blue, blotched, or speckled. Everything else within the aleurone, except the embryo, is endosperm, stored food for the embryo. The endosperm carries, like the aleurone, two sets of chromosomes from the female and one set from the male parent. The colors of the endosperm are less varied; it is usually colorless, or has yellows or oranges of several intensities, but there are several arrangements of the starchy contents. Colors are most apparent in hard corneous starch, which makes up most of the endosperm of pop, the outer layer of flint, and the sidewalls of dent. The colors are scarcely visible in flour corn, which appears white (or colorless) but may carry genes for yellow endosperm. An ear of corn on a plant grown from white flour seed could have some kernels that are light blue and slightly flinty if it had received some pollen from a blue flint corn.

The genetics of color and texture of grain are complex (Brown, Anderson, and Tuchawena 1952; Neuffer, Jones, and Zuber 1968) because the characters not only exhibit variable dominance, but also may interact, be concealed by inhibitors, and be present as one, two, or three sets depending on which parents contributed them. Color is a valuable indicator of purity of a harvest and most Indians distinguish their many kinds of corn by color, grain texture, size of ear, and length of growing season. Descriptions of corn grown by the Hopi (Whiting 1939, about 20 kinds), Zuni (Robbins, Harrington, and Freire-Marreco 1916; Bohrer 1960, more than 7 kinds), Omahas (Will and Hyde 1968, 13 kinds), and many other groups show this. For many Indians, colors in corn have "ceremonial" importance. The colors are associated with directions (Robbins, Harrington, and Freire-Marreco 1916; Whiting 1939).

Careful selection of seed to definite standards is essential if distinct kinds of corn are to be maintained. More than 250 mutants with visible effects have been described (Neuffer, Jones, and Zuber 1968) and many others have not been studied. Many mutants, like variations in disease resistance, day length factors, and response to fertilizers, are not visible in plants and must be studied in field experiments. Corn crosses freely, so farmers must select seed and plant it in separate fields or at different times to maintain the kinds they recognize.

Plants as Indicators of (Past) History

The plants people use are a key to the past and to their activities and environment at the time the plants were collected. Of all the plants used, cultivated ones provide the best evidence of human development. A cultivated plant and the techniques developed to grow it usually spread from a definite center. During the time a plant is being grown and spread by human activity, the plant is being affected and channeled by mutation, environmental selection, hybridization with wild and weedy plants and with other selections of the same species, by seed selection, seed storage techniques, and other factors. Thus, a farmer's harvest always differs from the seed originally planted. Cultivated plants become dependent upon people, and human life patterns often are governed by attachment to crops.

The corn ear is uniquely useful for comparative studies because it is durable, contains so many measurable parts, and is often brought back to the dwelling site. The volume of dried corncobs from some North American cave dwellings, or the carbonized ears from some open sites, is probably greater than the archaeological remains of wheat from all of Europe.

Corn, Culture, and Classification

If corn could grow without human help, or if people planted any seed harvested, then there would be a pattern of geographical variations in corn similar to that in many wild plants. Variation would be largely masked by dominant characters, short-season, and long-day variants in the north; heat-tolerant, long-season, and short-day kinds in equatorial regions; and variants adapted to dry or wet regions, and to high mountains with their cold nights. We can see such adaptations in the corn the American Indians grow. Good examples are the small and short-season eight-rowed corns of New England and adjacent Canada and the very short and deep-rooted plants grown in Arizona by the Hopi and Navajo. Superimposed on the regional adaptations, and obscuring them, are characters preserved through ideas of seed selection and methods of farming and seed storage.

Most Indians recognize and grow several kinds of corn, and know that they mix in their fields but attempt to maintain pure lines. The names they use, such as those in Whiting's (1939) list of twenty Hopi kinds, must be adequately qualified as in "James Jones' selection of Walpi blue flint corn" in order to be roughly equivalent to the category variety (or cultivar) used by horticulturists. Native names are useful and tell much about seed selection standards, but they are unavailable for archaeological remains. A system that can be used to make comparisons over a large area and through many years of evolution is needed.

It is significant that the major recognition characters used by Indians, those of kernel color and texture, are ones that may be modified by the pollen parent. The farmer, then, can usually select for seed those ears that have not been obviously contaminated by stray pollen, blown from fields of other kinds. The major kinds of corn recognized and grown by Pima and Papago of northern Mexico and southern Arizona and related peoples along the lower Colorado River, or by Hopi, Zuni, and Pueblo of northern Arizona and New Mexico, and those of all the agricultural Indians across the Plains to the Iroquois in New York were white, yellow, and blue corn, usually with flour and flint variants, and occasionally sweet. Bright red corn and variegated or striped variants, with the color carried in the outer layer of the kernel, which is derived from the ear-bearing parent alone, are far less common.

Classification systems developed by corn breeders of the past century followed the systems of the Indians, which were usually based on the character of the storage materials of the kernels and similar characters (Sturtevant 1899). These are artificial systems, designed to provide names for recognizable groups, and are still useful as descriptions of seed to be used for food or feed. With such systems, closely related kinds, such as the flint and flour variants of an Iroquois white corn, frequently are placed in widely separated groups.

More desirable for the study of the evolution of corn and its relation to variables in culture history is a phylogenetic system, one designed to represent supposed evolutionary patterns (Anderson and Cutler 1942). Such a system is complex. We have no simple methods to describe or graphically represent the many crosses and back-crosses, the introduction of variability through mutations and hybridization with wild relatives, the effects of inbreeding by small, isolated groups of people, and the effects of mixing of seed brought in by migrations, trade, war, and similar contacts. Carter and Anderson (1945) demonstrated gradual change in corn from the eastern Rio Grande Pueblos to the Hopi and refer to some effects of cultural change. Such changes, linking corn and culture, have been going on for a long time. An example is the dramatic change that occurred in the Mogollon area of western New Mexico during the Georgetown phase, about A.D. 500 to 700 (Cutler 1952:468). There had been a slow increase in the proportion of eight-rowed ears and similar decrease in the twelve-rowed ears used at Tularosa Cave during the previous thousand years of occupation, but between A.D. 500 and 700 the proportion of eight-rowed ears nearly tripled, from around 20 percent to 60 percent, while the twelve-rowed ears dropped from around 40 percent to 10 percent. This is an extreme example, but others can be found in archaeological and present-day materials (Cutler and Blake 1976).

Races of Corn

Major recognizable groups of corn are usually called races, in an informal system suggested by attempts to classify humankind. A race is defined as "a group of related individuals with enough characteristics in common to permit their recognition as a group" (Anderson and Cutler 1942). *Races of Maize in Mexico* (Wellhausen et al. 1952) contains photographs of Mexican races, uses numerous characters to demonstrate probable evolutionary patterns, and demonstrates reconstruction of some races by experimental hybridization of presumed parent stocks.

Most corn is a mixture of two or more races, but racial elements and possible relationships and lines of evolution can usually be recognized as long as the

Indian farmer maintains his traditional corn and grows it in fields away from introduced races. Hopi corn is slowly changing, but many of the kinds they grow are similar to those from nearby archaeological sites. Some farmers of the Rio Grande Pueblos are now growing mixtures of their old kinds of corn with modern Corn Belt dents. The greatest amount of mixing I have seen is in some collections that geographer Campbell Pennington (1969) made in northern Mexico. Corn grown by Indians in remote villages is similar to that found in archaeological sites, but Indians who live in larger towns and have frequent contacts with people from central Mexico grow mixtures of many kinds of corn, and their traditional kinds of corn are scarcely recognizable and are little grown.

Corn in North America

North American corn may be divided into four groups:

I. Small cob corn, with small, cylindrical, or cigar-shaped ears tightly wrapped in husks, generally with twelve or fourteen rows of small hard kernels: (1) Small Cob; (2) Chapalote; (3) Reventador; (4) North American Pop.

II. Corn with larger ears and kernels that probably evolved through continued selection of larger plants and kernels in a search for greater yield: (5) Pima-Papago, which has subraces Onaveño, Blando, and Western Eight Row; (6) Eastern Eight Row; (7) Midwest Twelve Row.

III. Dent corn and its many mixtures: (8) Toluca Pop; (9) Conical Dent; (10) Pueblo; (11) Cristalina; (12) Corn Belt Dent.

IV. (13) Sweet corn.

Races of Corn

1. *Small Cob.* The most ancient corn from archaeological sites in northern Mexico and the Southwest has small ears with long glumes that partially cover small hard kernels. The cobs are similar to some from the Tehuacán caves of east-central Mexico that Mangelsdorf, MacNeish, and Galinat (1967b) call wild-type segregates and similar to a few cobs from Tamaulipas caves, which they call Pre-Chapalote (Mangelsdorf, MacNeish, and Galinat 1967a). The Tehuacán cobs may date from 5000 B.C.; the cobs from Tamaulipas in northeastern Mexico from about 2300 B.C. Occasional eight-rowed ears are most primitive and probably oldest, but twelve- and fourteen-rowed ears are common in the early levels. Small Cob has been found in Bat and Tularosa caves of west-central New Mexico and in caves dated as late as A.D. 1200 in Hidalgo County, New Mexico, and A.D. 1300 in Red Bow Cliff Dwelling (Arizona W: 9: 72) on the San Carlos Apache Reservation, Arizona (Cutler 1965a). A few Chapalote ears that Carl Sauer collected in northern Mexico resemble Small Cob and suggest that this primitive race may persist in remote areas.

2. *Chapalote* appears to be a hybrid between Small Cob and teosinte. It is a primitive race of hard pop or flint with small cigar-shaped ears. Its twelve or fourteen rows of kernels have smoky brown pericarp visibly striated by tight, brittle husks. Glumes are long and several male flowers frequently are borne on the tip of an ear. Chapalote is mainly found in northwestern Mexico and rarely north of the Papago country.

From the small ears of Small Cob, there is a progression through Chapalote, Reventador, and Pima-Papago, to Eastern Eight Row with its eight rows of large grains. The limits of each race are artificial, maintained by human selection but useful in descriptions.

3. *Reventador* has flatter kernels, usually has more rows of grains, and generally is larger than Chapalote. *Reventador* means "popcorn" but many ears are hard flint. It apparently developed from Small Cob and Toluca Pop and is still a greatly variable race in northern Mexico and the Papago country of southern Arizona.

4. *North American Pop* is a loose category that includes nearly all the pop and very hard small flint north and east of southern Arizona. This race is derived from Small Cob, Chapalote, and Reventador. Archaeological specimens have sometimes been called "Tropical Flint." The race includes the earliest small corn from Adena and Hopewell sites (Cutler and Blake 1976), short-season and eight- to twelve-rowed pop from the Atlantic coastal area (Sturtevant 1899), and the many variants of indigenous pop grown by Eastern and Plains Indians up to recent times (Will and Hyde 1968).

In the past few years there have been changes in North American Pop as a result of mixing with plants grown from seed purchased from catalogs and seed stores. We have seen harvests at several Rio Grande Pueblos and Hopi villages that included ears of Japanese Hulless and South American Golden Pop grown from seed purchased from seed catalogs. On the roof of a house in Acoma in the fall of 1956, ears of a Reventador-like corn, which the owner called their "old corn," were drying side by side with "foreign" popcorns. Most ears showed in color and shape of grains and ear the effects of crossing.

5. *Pima-Papago* (Anderson and Cutler 1942) has been the dominant race of corn in northern Mexico and the southwestern United States for nearly 2,000 years, and a form of it spread eastward through Oklahoma and Arkansas and up the Mississippi River to reach northern Illinois by about A.D. 100, and southern Georgia perhaps even earlier. Pima-Papago apparently evolved from the small pop and flint races through selection of variants with large ears and kernels. The oldest forms merge into Reventador and have twelve or fourteen rows of flinty kernels that are only slightly wider than they are thick. Ears of this type have been called Onaveño (Wellhausen et al. 1952; Cutler 1965a). Ears with ten or twelve rows of slightly larger kernels, often flour but in prehistoric specimens usually flint, have been called Blando de Sonora (Wellhausen et al.

1952) or Blando (Cutler 1965a). Eight-rowed forms of Pima-Papago have been considered a race, Harinoso de Ocho (Wellhausen et al. 1952) or Maiz de Ocho (Galinat 1965). We consider Western Eight Row a subrace merging into Blando and often distinguishable only by the number of rows of grains. A single plant may bear ears with eight and ten rows of grains. It is difficult to separate the subraces of Pima-Papago, when specimens consist only of cobs recovered from prehistoric sites.

Pima-Papago ears are usually white or yellow but frequently blue, and there is a wide range of other colors as well as variegated (or calico), dotted, speckled, blotched, and other markings. The hard and slender forms of blue corn and of the deep purple-black corn of the Pueblo, Zuni, and Hopi Indians of Arizona and New Mexico, with twelve or fourteen rows of hard flint kernels, belong to the Onaveño subrace. Kernels found in archaeological sites usually are flint. The hardness, which protects stores of food and seed but also made food preparation difficult, is scarcely needed today when tight storage vessels are available and danger from food shortages diminished. Hopi informants say that they now grow less flint corn (Whiting 1939), and scarcity of flour corn in nearby archaeological sites confirms this trend from less flour/more flint to the reverse proportions (Cutler 1965a).

A cache of 222 entire ears from a Basketmaker level of Mummy Cave in Canyon de Chelly, Arizona (Anderson and Blanchard 1952), contained mostly twelve-rowed flour ears, similar to ears still grown by many Papago and to the traditional corn, now seldom grown, of the Pima, Cocopa, and Maricopa. It was largely this collection, now at the American Museum of Natural History, that caused Anderson and others to refer to this type as Basketmaker corn, even though similar corn is found in earlier and later sites in the Southwest and is still grown. The Hopi still grow some almost pure strains of this Pima-Papago race. It is the major element in their most common corn, a large, white, floury corn with a big shank and occasionally dented grains that we classify as the Pueblo race of corn.

The Western Eight Row subrace became the dominant corn of parts of Arizona after about A.D. 500. It has been suggested that Western Eight Row came from South America (Wellhausen et al. 1952), but it is more likely that selection for larger grains and natural selection for smaller plants adapted to short seasons and severe environmental conditions resulted in eight-rowed corn. The Georgetown phase (A.D. 500–700) levels of Tularosa Cave, which contained more eight-rowed cobs than other levels, also contained a large proportion of wild plant remains (Cutler 1952), an indication that conditions for agriculture were marginal, favoring natural selection for eight-rowed plants.

Western Eight Row spread northward beyond central Arizona and New Mexico but did not become the dominant corn there, perhaps because farmers selected strains with more rows or because there was some hybridization with multirowed dent corn.

6. *Eastern Eight Row* was described as Northern Flint (Anderson and Cutler 1942; Brown and Anderson 1947), but it was later found in recent and archaeological collections over most of the United States east of the Rockies (Cutler and Blake 1976). Eastern Eight Row reached as far north and east as Ontario, Canada, before A.D. 800. By A.D. 1200 Eastern Eight Row dominated most of the region east of the Mississippi and by A.D. 1500 covered most of the region east of the Rockies. Indians and Europeans moving westward continued to spread it. Of 1,714 cobs from Sheep Rock Shelter, Pennsylvania, a site dated about A.D. 1550, 88 percent were eight-rowed and most of these had heavy shanks and hard cobs. Only 11 percent were ten-rowed. In the Plains and along the Gulf Coast, Eastern Eight Row includes more ten- and twelve-rowed ears, usually about 30 percent at the time of European contact. Iroquois of New York, some Cherokee in South Carolina, and Fox at Tama, Iowa, still grow relatively pure forms of Eastern Eight Row. Golden Bantam sweet corn is derived from an Eastern Eight Row corn carrying the sweet gene. Eastern Eight Row and Midwest Twelve Row (see next race) arose from eastern expansions of the Pima-Papago race and differ from it mainly in their harder cobs and glumes and in their shorter and wider kernels. Eastern Eight Row continues to be important because crosses of it with dent corn produced the Corn Belt Dent upon which most commercial corn production in North America is based (Brown and Anderson 1948).

7. *Midwest Twelve Row* is almost extinct. It spread to the east as a variant of Pima-Papago and persisted in the central and lower Mississippi Valley until it was largely replaced by Eastern Eight Row (Cutler and Blake 1976). Occasional ten- and twelve-rowed ears found in the corn of some southern and Plains areas may be relicts of the disappearing Midwest Twelve Row.

8. *Toluca Pop* is an ancient, small, and many-rowed pop, flint, or dent, which is an ancestor of the common dent races (Wellhausen et al. 1952). It is rare in a pure form in northern Mexico either in archaeological sites or among Indian farmers (Pennington 1969). Toluca Pop has a short, conical ear with long, slender, often pointed kernels. It is often pop but frequently dented and shows fasciation or distortion that multiplies the rows of grains. In historic times Toluca Pop was brought north from Mexico and widely grown as popcorn. Some of the common cultivars are Baby Rice, Japanese Hulless, and the ornamental Strawberry Pop.

9. *Conical Dent* is largely derived from Toluca Pop (Wellhausen et al. 1952). Under this term we include all the Mexican Pyramidal, Fremont Dent, Conico, Zapalote Chico, Pepitilla, and similar races of extreme northern Mexico and the southern United States, with the exception of the locally evolved races, Pueblo, Cristalina, and Corn Belt Dent. The earliest dents we have seen from northern Mexico are from a cave on the Rio Zape, Durango, Mexico, whose oldest levels

are dated about A.D. 660. The most extreme dents, and the earliest ones, from north of Mexico are from Fremont culture sites (Cutler 1966), which may date about A.D. 1050 to 1200. Some Fremont ears resemble Conico and Zapalote Chico of Mexico (Wellhausen et al. 1952), and rare ears approach the extreme denting of Pepitilla. These prehistoric dents are important because they brought another set of characters, ones that were compatible with the very different set in the Pima-Papago race, and combined to produce new races, called Pueblo and Cristalina, described below. These races did not spread beyond parts of the Southwest.

Historical records for dent corn north of Mexico begin about 1678 (Ewan and Ewan 1970) when dent corn was being grown in Virginia by Europeans while local Indians grew other kinds. Dents were found in Texas in eighteenth-century Spanish mission sites (Cutler and Blake 1976). Most of the dents grown in the southern United States today can be traced to post–A.D. 1500 introductions from Mexico. Brown and Anderson (1948) describe these as a unit they call the Southern Dents, and the group includes Tuxpeño, Pepitilla, Olotillo, and several other of the Mexican dents described by Wellhausen et al. (1952).

Today Indians in the United States grow some dent corn, and these dents have mixed with their traditional corn. In northern Mexico and among the Pima and the Papago who use irrigation, the most common corn is the twelve- to sixteen-rowed white Tuxpeño.

By the mid-1800s it was known that plants from mixtures of Southern Dents and Eastern Eight Row had superior qualities. It was from these mixtures that modern commercial corn was selected (Anderson and Brown 1952).

10. The *Pueblo* race of corn (Anderson and Cutler 1942; Carter and Anderson 1945; Cutler 1966) has been found in sites in Glen Canyon and at Betatakin, Inscription House, and a few other post–A.D. 1050 sites in northern Arizona. It is now grown by Hopi and Zuni and is the most important corn of the Pueblo, Navajo, and Apache. Until recently we considered prehistoric dent influence to be limited and of little importance. Corn from several levels of Picuris (or San Lorenzo Pueblo), excavated by Herbert Dick, and of Gran Quivera, New Mexico, excavated by Alden Hayes (Cutler and Blake 1976), suggested that the most dented forms of Pueblo corn in eastern New Mexico appeared after the Pueblo Rebellion when settlers returned from Mexico bringing dent corn seed. Extreme dents of some Rio Grande Pueblos, especially the large rust-colored dents of Acoma and Jemez, may be even more recent.

The wide range of dent corn from a cache in Antelope House, Canyon de Chelly, recently excavated by Don Morris, and from other northern Arizona sites, suggests that dent corn contributed hybrid vigor to early Pueblo III corn in this region and some of this spread to the Rio Grande. Dent corn is common in Fremont culture sites and seems to spread from that region, although it is also possible that dents came directly and independently from Mexico to the

Pueblo as well as Fremont cultures. That this corn did not spread farther and have more effect on prehistoric corn production may have been a result of the drought and extensive migrations of the late 1200s.

The gradient that Carter and Anderson (1945; Cutler 1966) found from dented, larger, more "Puebloid" corn of the eastern Pueblos, to the more Pima-Papago–like corn of the Hopi and Zuni is apparently a result of post–Pueblo Revolution introduction of more dent corn to the Rio Grande settlements. Archaeological materials show an earlier gradient in the opposite direction, with more denting near the Colorado River and very little at the eastern Pueblos.

Modern Navajo corn in collections at the Missouri Botanical Garden, even those collections from Shonto, not far from the Hopi villages, resembles corn from the eastern Pueblos, especially Zia, Jemez, and Acoma, far more than it resembles corn grown by the Hopi. We would expect this because the Navajos' early agricultural contacts were with the eastern Pueblos.

11. *Cristalina* (Wellhausen et al. 1952) is, like the Pueblo race, a mixture of Pima-Papago corn with dent. It is a vigorous race, the dominant corn of at least part of northwestern Mexico by A.D. 600 (Brooks et al. 1962), and it is still grown by Indian farmers in that region (Pennington 1969).

12. *Corn Belt Dent,* like the Pueblo race, resulted from a cross of a large-grained and few-rowed local corn with the many-rowed dent from Mexico. Corn Belt Dent appeared at a favorable time and has spread so that, according to the United States Department of Agriculture, the 1973 production from the United States alone was over five and a half billion bushels, with the average yield per acre over 90 bushels. It is so widely grown that in most places it has supplanted old Indian varieties.

Corn Belt Dent was selected from a relatively small number of collections, and the inbreeding practiced by corn breeders to increase yield and uniformity has reduced the variability. Further selection to produce types that can be manipulated to yield uniform seed for marketing has eliminated more of the diversity that characterized the crop grown by the American Indians. There is great danger that pests or diseases may arise that could destroy much of the harvest (Miller 1973). In 1970, corn leaf blight destroyed about 15 percent of the crop. The many kinds of corn still being grown by American Indians form a reserve of germ plasm for corn improvement and protection that must be preserved.

13. *Sweet corn* has wrinkled and translucent kernels, the effect of a recessive gene that makes the endosperm sweet. The most extreme types, with many rows, often more than twenty, of usually golden yellow or red, slender kernels on a conical or pineapple-shaped ear, are found in highland valleys of Bolivia and Peru. Their glumes are soft and small, the leaves a paler green than leaves of other kinds of corn. There is a gradient from these extremes of western South America to the straight and eight-rowed Golden Bantam ears of the

northeastern states. The sweet corn of any region usually combines characters for rounded ear shape, soft, small, white glumes, and many rows of frequently reddish or yellow kernels with the characters of the local corn. Thus, sweet corn of the Plains Indians tends to have ten to sixteen rows of often reddish grains, while other kinds of Plains Indian corn usually have eight or ten rows of slightly broader and shorter kernels. Sweet corn, never very important among the Indians of North America, was rarely eaten as green corn on the cob (Whiting 1939; Will and Hyde 1968; Wilson 1977). Travelers' references to "green corn" and "sweet corn" usually refer to slightly immature corn, picked when the kernels are still moist and doughy but riper than the milk state, which is preferred by modern eaters of green sweet corn or even for roasting ears. Some Papago, Hopi, Zuni, and Pueblo Indians still cook green corn ears and dry them for later use, a practice that was common on the Plains (Wilson 1977). Most of the translucent and slightly shriveled kernels we have seen from archaeological sites are kernels from baked or steamed green corn. Only a few are sweet corn.

The sweet corn gene is recessive. If kernels on a plant of pure sweet corn are pollinated by pollen from a plant that is not sweet, then the kernels will not be shriveled and translucent when dry as are those of sweet corn, and would not be picked for sweet corn seed. Thus, the cluster of characters associated with sweet corn could be preserved for a long time by Indian farmers exercising their usual care in seed selection.

Corn and Social Structure

Most Indian families grow several to many kinds of corn and select and maintain their own seed (Whiting 1939). In 1953, during our survey of corn in the Southwest, a woman at Acoma told us, as we examined her rooftop of drying corn, "The hard yellow corn my neighbor grows is harder than mine." Other Indians mentioned differences in corn grown by different families. Family "strains" of corn apparently exist preserved by the separation of fields, by individual seed selection standards, and by the rarity of seed exchanges (Robbins, Harrington, and Freire-Marreco 1916).

Corn pollen is heavy, the largest pollen in the grass family, and does not travel far. Most pollen falls on nearby plants, usually within the family field. These fields often adjoin fields of other families, usually related ones or ones belonging to the same clan (diagrams of field locations may be found in Wilson 1977; Forde 1963; Bradfield 1971). Wilson's (1977) Hidatsa friend, *Maxí diwaic*, said that her family often agreed with families using adjacent plots to plant the same kind of corn. Cross-pollination in this case could not be detected by color or grain texture, so some mixing of family strains could be preserved. Clans or other divisions of a settlement usually planted in separate areas, so there was relatively little cross-pollination outside the clan. Fields of different communities, settlements, or villages were even more isolated. Frequency of interchange of crop plants and cross-pollination is dependent on social interchange. The system is comparable to that of small islands (families) in an archipelago (clan or similar unit) in a sea (community) with other archipelagoes. Similarities in corn should be directly proportional to closeness of human relationships. Comparisons of the corn of families, or of related groups of clans within a settlement, are complicated by the variation within each of the several kinds of corn grown and by the cultural changes that are affecting native crops, but can be made on large collections of fresh ears.

Corn from archaeological sites usually consists of carbonized material or cobs without kernels. The kernel color and texture and ear shape by which farmers identify their crop cannot be seen, but there are other characters to measure or count. The number of rows of kernels may be counted by the spaces left on a cob, or estimated by measuring the angles of the sides of a single kernel and dividing this into 360 degrees. (For example, the sides of a kernel from an eight-rowed ear make an angle of 45 degrees, which is one-eighth of a full circle.) The thickness of a kernel can be measured directly or by the space it left on a cob. Measurements of the size and shape of the many parts of a cob can be used for comparisons. The cupule, a socket in the cob that lies beneath a pair of kernels, varies, from the very small, delicate, and open one of most popcorn to the extremely large, hardened, and compressed cupule of Eastern Eight Row (Nickerson 1953).

Various methods have been used to compare corn from different communities, regions, or time periods within a site. Longley (1938), in his study of contemporary North American Indian corn, found that plants from the most northern Indians had fewest knobs, or deep staining regions, on their chromosomes, and that most knobs were found on plants in his collection most like corn of northern Mexico. Carter and Anderson (1945) used graphs to show that the corn of the eastern Pueblos showed more Mexican influence than corn of the Zuni and Hopi. For comparisons of the kinds of Mexican corn, Anderson (1946) used pictorialized scatter diagrams. So did Nickerson (1953) when he studied twenty-three lots of corn from a wide range of peoples and times in a search for characters that could be used to identify corn types. Brown et al. (1952) used fifteen cob and tassel characters and observations on chromosome knobs to compare three varieties of Hopi corn. In our studies (Cutler 1960, 1966; Cutler and Blake 1976), we have tried to use the most obvious and simple measurements, ones that can be found easily on modern or archaeological specimens, in an attempt to encourage more people to collect and study corn, and to speed the examination of large collections. These relatively simple methods reveal major patterns and can be extended by the use of more characters.

Corn Growing and Storage

Throughout the Americas, patterns of corn growing are surprisingly uniform. Usually four to eight kernels of corn are planted in a hole; as the plant grows, soil is heaped around it into a hill. Holes are spaced about 3 or 4 feet apart, although they may be closer in good soils or as much as 8 or 10 feet apart in the drier parts of the Southwest. There are many variations on this pattern. In the northern states, where soils are wet, hills may be used repeatedly and the seed planted on the slope. In the dry Southwest, seeds may be planted at the bottom of a 12-inch-deep hole and soil gradually added about the growing plant.

At harvest time, the ear, in its husk, was usually removed from the plant in the field. Harvesting and husking corn frequently were occasions for social gatherings. Sometimes all of the husks were removed. Often about four husks were left and the ears braided into strings that could be hung on racks, in the house, or on the roof to dry (Wilson 1977; Waugh 1916; Castetter and Bell 1942). We have seen carefully stacked ears with only a few husks on them in a storage room at the Hopi village of Moencopi in 1953 and in a well-preserved burned corn crib of about A.D. 1300 in southeastern Missouri (Cutler and Blake 1976).

Seed corn was usually selected at harvest time and stored as entire ears. Many Indians stored enough seed for two plantings, generally planting the oldest seed first.

Plains Indians and those to the northeast often stored corn in large, carefully prepared underground pits, and occasionally in above-ground storage structures, or in parts of the living quarters (Waugh 1916; Wilson 1977; Driver and Massey 1957). In the Southeast, special storehouses, usually raised above the ground, were used. In the Southwest, storage was in special rooms in the dwelling complex, sometimes in cache pits or buried pots or baskets, or in structures built under rock overhangs. The Papago and others in southern Arizona and adjacent Mexico usually built coiled grass or twig basket-like structures on their roofs or on platforms near their houses (Castetter and Bell 1942, 1951), or built adobe granaries under rock overhangs.

Uses of Corn

Nearly all of the many uses of corn are widespread. Preferred food preparation methods vary from one region to another, but the most striking feature is their similarity throughout the New World. Champlain, Sagard, and other early travelers among the Huron mention their preparation of "stinking corn." It is prepared by soaking ears of fresh corn under water for two or three months and cooking it. Modern Iroquois do not know this food (Waugh 1916). Rick and Anderson (1949) report a similar preparation from the Sierra de Ancash in highland Peru. There, as long ago among the Hurons, fresh ears are soaked for two or three months and the corn cooked before eating. About 20 percent of the corn eaten in the Sierra de Ancash is prepared in this fashion.

Toasted, or parched, corn is probably one of the oldest foods. About half the corn eaten in the Sierra de Ancash is parched. Among the Tepehuan of northern Mexico, the two most common foods are parched corn (Pennington 1969): for *pinole*, grains are toasted and ground fine, sometimes with flavoring leaves; for *esquiate*, the toasted grains are mashed with water and flavoring to prepare a paste. Parched corn was prepared by the Plains Indians (Wilson 1977) and by the Iroquois (Waugh 1916), who use it now only in ceremonies. Most corn-growing North American Indians carried parched and ground corn on trips, boiled balls of moistened meal in water to prepare dumplings, and wrapped mush in inner husks and boiled, steamed, or baked their variants of the tamale.

Many of the utensils, materials, and methods of preparing corn are ancient and widespread ones. Ashes, for example, were widely used to make acorns less bitter (Waugh 1916; Driver and Massey 1957). Most Indians soaked corn grains in ashes to remove the tough outer skin, the pericarp, and to soften and swell the grain. Tortillas in Mexico were made from finely ground ash-treated grains, although today lye is often substituted for ashes, or prepared tortilla dough is purchased. Ashes also keep the batter used by the Hopi and Zuni for their thin blue *piki* bread alkaline and blue (Stevenson 1915).

It is generally thought that alcoholic beverages made from corn spread only to northern Mexico and barely into the United States (Driver and Massey 1957). The many restrictions on alcohol production have made it difficult to secure reliable information. Pennington (1969) records methods and many plant additives used for corn brewing among the Tepehuan, which indicate that techniques were well developed in northern Mexico. The first step in brewing corn is to convert some of the starch to sugar in order to feed the yeast, which, in the second and final step, will produce alcohol and the carbon dioxide that gives well-made corn beer its sparkle and acidic tang. Starches can be changed to sugars by the addition of enzymes present in sprouting grain and in saliva. Both sources were used by New World Indians.

Today most corn beer, the *tesquiño* of northern Mexico or the *chicha* of South America, is prepared from sprouted corn, but in the past much was prepared from salivated corn meal. The Pueblo and Zuni do not prepare corn beer today, and there is no archaeological evidence that they did in the past, but they have been reported to use salivated corn meal to prepare a sweet baked meal, and to grind sprouted and dried corn, which they mix with water for a sweet drink (Stevenson 1915; Robbins, Harrington, and Freire-Marreco 1916). The Zuni also use saliva to make a conserve of boiled yucca pods sweeter (Stevenson 1915), a technique that seems to be ancient.

Wheat flour has taken the place of corn for many Indians, and each year fewer Indian farmers plant the many special kinds of corn they formerly used. Each year some of the family strains, which developed through years of selection, are lost. Some Indian corn has been collected and is kept in storage at the U.S. National Seed Laboratory, but the number of collections and the amount of seed in each one are so small that this represents only a minute fraction of the great diversity still apparent in Indian corn grown today (Miller 1973). Corn breeders have found some genes for more and better kinds of proteins and for resistance to changing diseases and insects, but many other useful genes might be extracted from the largely untapped resources of corn being grown today by the American Indians.

2 Cultivated Plants from Picuris

Hugh C. Cutler
Missouri Botanical Garden and Washington University

(This work was aided by National Science Foundation Grant G17593, 1966)

Leonard Blake's Comments, 1999
　　In the early 1960s, Dr. Herbert Dick was asked by the Picuris Indians living in the Pueblo of San Lorenzo in northern New Mexico, south of the better-known Pueblo of Taos, to aid them in getting a water line into the Pueblo. Dr. Dick laid out and excavated the line, and, in doing so, collected archaeological specimens, which included pottery and plant remains. Because variations in the pottery from archaeological sites in the southwestern United States are now so well known, it is commonly possible to use the kinds of pottery found to date other associated remains. This was the case at Picuris, so that Cutler was able to demonstrate the changes that had taken place in the kinds of corn grown in recent years, after the Pueblo revolt of the seventeenth century. Cutler also discusses changes in corn in the Southwest over earlier periods. A number of the comments made about "North American Indian Corn" (Chapter 1) also apply to this report.

　　Cultivated plants may be considered as special kinds of artifacts. They are created, maintained, and transported by people. Most American Indians grow several different kinds of a plant. Within each kind there are many variants, the product of years of care by a family or similar group that grew its crops in separate fields and selected its own seed, and of environmental selection and chance hybridization. There is sufficient contact between the various Indian groups so that similar kinds of cultivated plants are grown over wide areas. Consistent differences in the crop plants usually can be related to the distance from Mexico, environmental and cultural complexity, and relations with neighboring people.
　　The ideal way to study corn or any other plant from living Indians or from archaeological sites is to compare each major kind of corn with its counterpart from other villages, sites, and time periods. Although this can be done with good contemporary material, our present techniques and abilities very seldom enable us to identify with certainty each kind or cultivar in the usual archaeological

collections. In nearly all cases we must compare entire lots or attempt to distinguish natural groupings. In the diagrams (Figures 2.1–2.7) some of these natural groupings of related kinds can be seen as clusters of points. These groupings usually include several closely related kinds of corn selected on the basis of color or texture by the growers. By careful comparisons of cultivated plants from a site with those from other sites we may be able to trace the relationships of the plants and the people who grew them. The long occupation of Picuris and other Pueblos and the relationship of the Pueblo people to those who occupied older sites make it possible to follow changes in their cultivated plants and to demonstrate how changes in cultivated plants are linked to other steps in the evolution of a culture.

The Corn Ear

The corn ear follows the general pattern for fruiting parts of most grasses, but it has been so compressed and hardened that some training and practice is needed before one can effectively measure and describe the parts.

Two of the most useful characters are relatively easy to see and record. The rows of grains can nearly always be counted. Even when the grains and the spikelets (the chaff of the cob) that bore the grains are missing, it is possible to count the cupules in which each pair of spikelets with its pair of grains was borne.

Measurements of the gross diameter of cobs from archaeological sites usually are inaccurate because varying quantities of the spikelets are worn off. In addition, the number of rows of grains and the lengths of the spikelets affect the diameter. Nickerson (1953) discovered that the width of the cupule in which a pair of spikelets is borne is a reliable measure of the size of the central axis of the cob.

Cupules from the tip or butt of an ear or from distorted or unusual rows should not be used, but even fragments including a single cupule often can be used. Width is measured across cupules from one margin to the other at right angles to the longitudinal axis of the ear. In general, there has been a reduction in the number of rows of grains in Southwestern corn during the past 2,000 years (Martin et al. 1952) and an increase in the size of the cob. The most important exceptions to this trend are found in the corn of the eastern Pueblos, the corn from the Fremont culture sites in Utah and Colorado, and corn from a few sites in the lower Mississippi Valley.

At the present time there is no adequate collection of corn with data on seed selection, planting, growing, harvest, number of kinds grown, specific uses, and the many relations to ordinary and ceremonial life, legends, and trade, for any living or dead group of American Indians. This is also true for squashes, gourds, and every other cultivated plant and for most of the wild

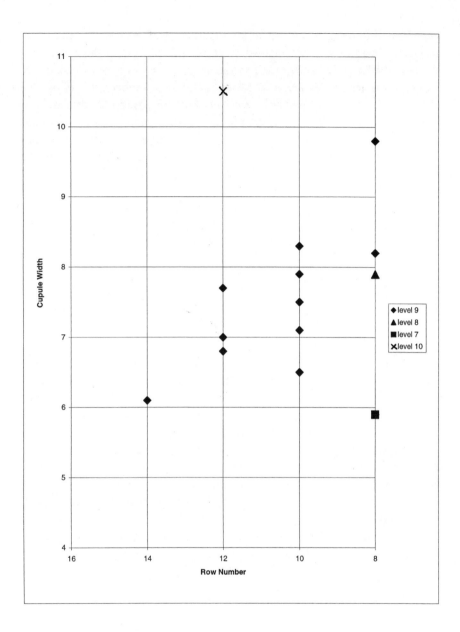

Figure 2.1. Corn cobs from the Taos phase, A.D. 1150–1225. TAIII, Test Pit O. (All measurements on carbonized cobs are corrected for an estimated 20 percent shrinkage in cupule width.) Level 7 = ■; Level 8 = ▲; Level 9 = ♦; Level 10 = ×.

ones. The scattered published material usually has scant value because it is not supported by museum specimens or adequate illustrations. Useful collections are described briefly in Whiting's *Ethnobotany of the Hopi* (1939) made by Volney H. Jones and Alfred F. Whiting, mainly in October 1935, as part of their Hopi Crops Survey and are at the Museum of Northern Arizona and the Ethnobotanical Laboratory of the University of Michigan. About a third of the 125 Pueblo ears Carter and Anderson (1945) measured are in the collections of the Missouri Botanical Garden. Measurements and some specimens from a survey of Hopi white flour, blue flour, and purple (or Kokoma) corn made by Brown, Anderson, and Tuchawena (1952) are also at the Garden. The senior author made a large collection of Pueblo corn in October 1953 for the National Academy of Sciences-National Research Council Committee on Preservation of Indigenous Strains of Maize (Clark 1956). Although this is the largest collection available here, it lacks native names, selection criteria, and other data essential to a thorough understanding of the relationship of the corn to the culture.

Corn in the Southwest

The oldest kinds of corn in the Southwest are small-cobbed, usually twelve- to fourteen-rowed, with flexible glumes. This corn has been found in Bat Cave (Dick 1965) and Tularosa Cave (Martin et al. 1952), but apparently it was still grown in the Southwest as late as A.D. 1100 in southern New Mexico and Arizona (Cutler 1965b; Cutler and Meyer 1965). Related races, Chapalote and Reventador, are found in many sites and are still grown in northern Mexico and rarely by Indians in southern Arizona. By A.D. 700 most corn in the United States was derived from crosses of a large-grained, eight-rowed flint or flour corn with an older, small-cob, hard flint corn. (For a discussion of the kinds of corn grains see Cutler 1966:7–9, 43.) In the Southwest most of the older intermediates were twelve-rowed and belonged to the variable race called Pima-Papago (Anderson and Cutler 1942). The small-grained, flinty forms closest to Reventador are called Onaveño (Wellhausen et al. 1952). The larger-grained, floury forms, with fewer rows of grains than Onaveño, belong to a race called Harinoso de Ocho by Wellhausen et al. (1952), or Eight-rowed Maize by Galinat and Gunnerson (1961). This corn may also have flint grains; it is related to and not greatly different from much of the corn of the Plains and east of the Mississippi. Corn of the northeastern states was predominantly eight-rowed and hard-cobbed, perhaps the product of large doses of Mexican Harinoso de Ocho, some crossing with a related grass, *Tripsacum dactyloides,* and environmental and human selection. Extensive changes in corn in the Southwest took place during the period from A.D. 500 to 700, the time of the Georgetown phase in the Mogollon area (Martin et al. 1952), and slightly later in northern Arizona, New Mexico, and adjacent Colorado (Cutler and Meyer 1965), probably during Pueblo II times (A.D. 900 to 1075).

The first dent corn in the Southwest probably came into northern Arizona and Utah about A.D. 700 and traveled northward, but in Utah was restricted to Fremont culture sites. This was a typical western Mexican Pyramidal Dent, usually strongly tapered and fourteen-rowed, similar to Conico Norteño, Zapalote Chico, Pepitilla, and other Mexican races derived from Toluca Pop. It has been found in several sites west of the Colorado River in Utah and as far north as the Yampa River Canyon in southwestern Colorado (Cutler 1966). We have not been able to determine how much this dent affected corn in the Southwest. There is no evidence that it went southeastward from the northern Fremont area to the upper San Juan River region and thence to the Rio Grande Pueblos to introduce denting.

Two major areas of cultivated plant assemblages in the Southwest were defined by Carter (1945) twenty years ago. One includes the desert and river region occupied by Indians like the Pima, Papago, Maricopa, and Yuma who grow the Pima-Papago race of corn, tepary beans, and common beans. The other is the area of the plateau peoples, like the Pueblo and Navajo, who grow the Pueblo race of corn, some Pima-Papago corn, common beans, and very few teparies. These areas overlap and the borders cannot be strictly defined, but there are consistent differences in the kinds of cultivated plants grown in recent times. These differences apparently were slight in the very early periods of agriculture and the limits of crop plants are imposed by environmental factors, which halted the plants in their movement northward from Mexico, and by migration and culture patterns. Because Picuris lies near the northeastern margin of the Pueblo area, plant remains recovered from it are valuable for defining sources and times of changes that took place in the plant assemblages of the eastern Pueblo region.

Carter and Anderson (1945) point out the gradual change in corn grown today as one proceeds from east to west through the Pueblos. Corn of the eastern Pueblos is most "Puebloid," the western more like that of the desert tribes of southwestern Arizona. Carter and Anderson combine data from several different kinds of corn grown in each Pueblo. A review of the specimens they studied and of collections I made in 1953 shows two kinds of differences between corn of eastern and western Pueblos. With few exceptions the Pueblos grow the same general kinds of corn. In the east, any single kind of corn, such as white flour corn or blue flour corn, usually has more rows of grains, has a larger cob, and is apt to have some flat and slightly dented kernels. The same kind of corn grown at Zuni and the Hopi villages will be more like the corn of the desert and river tribes.

In addition to the east-west gradient of characters in a single kind of corn, there is a change in the amounts of each kind grown. Some extreme kinds are found only near the ends of the distribution line. Thus, some small ears of soft flour corn of the Pima-Papago corn race are frequently grown at Zuni and the Hopi villages but rarely seen along the Rio Grande. A large-butted, many-rowed, rust-colored dent corn is most abundant at some Keresan Pueblos and at Jemez, and practically never seen among the Hopi.

Denting is rare in the eastern Pueblo region before at least A.D. 1350. Most archaeological specimens of grains that show a slight dimpling of the cap were immature at the time of harvest and the cap collapsed when the contents of the grain dried and shriveled. Flinty material, if present in these dimpled or dented kernels, is not restricted to the sides of the grains as it is in the true dents. Practically all of the Picuris material, except the modern ears, is carbonized. Consequently, color and character of the storage material cannot be distinguished readily but extreme denting, had it been present, could have been detected. A few grains from the Trampas phase probably were slightly dented.

Denting is usually associated with more rows than are found in flint and flour ears, with flattened grains, and with enlarged cobs and butts. About 12 percent of the cobs from the Trampas phase (A.D. 1600–1696) either were dent corn or were the progeny of mixtures with dent corn.

For comparison with the archaeological specimens we have forty-three Picuris ears that Herbert Dick obtained from Pat Martinez in 1963 and thirty-six ears I selected from the harvest of Tom Martinez ten years earlier. Neither is a random sample or a complete representation of Picuris corn, but the total is useful for comparison. Only one-fifth of the ears bear some slightly dented kernels. There are no extreme dents, yet all of the ears have enlarged butts, many rows of flattened kernels, and other characters derived from dents. At Picuris, near the eastern margin of the Pueblo region, and outside the active center of movement of the Pueblo people and the Spanish, the older kinds of corn persisted and at the same time there probably was some slight mixture with corn from the Indian groups to the east.

From the late appearance of dents at Picuris we might conclude that extreme dents were introduced by the Spanish into the Pueblo region. This is supported by the limited distribution of the most extreme dent of all, the large-cobbed, rust-colored dent mainly grown in the western Keresan Pueblos of Acoma, Laguna, and Mesita, in Jemez and Isleta, and occasionally in Santo Domingo, but rarely seen in other Rio Grande Pueblos (and Picuris), and, as far as I know, never seen in the Hopi villages. This rust-colored dent is similar to some corn of western Mexico, although flour and flint corn of similar color but of far different cob shape and row number has been collected in modern Pueblos and from archaeological sites. This corn, either in white or rust-colored form, was probably brought from Mexico some years after 1692, for many of the old Spanish-American settlements in the region do not grow it.

The fact that some of the existing Spanish-American settlements do not grow the rust-colored, large dent might suggest that the dents were brought into the region by trade among the Indians or movements of Apaches or other mobile groups. A linguistic study of varietal names and corn uses might suggest the source of Pueblo dents.

Some dents were present before the Spanish arrived, but none of these is extreme. There is evidence in crop plants grown by Indians in the Southwest that a series of changes were introduced from Mexico (Cutler and Whitaker 1961; Cutler 1966:3). At the time Europeans arrived in the Southwest the most common corn apparently was yellow flint, with white flint, yellow flour, and white flour corn apparently following in about that order. Some blue corn was grown, but it is difficult to determine how much because blue disappears with age. Red, pink, variegated, and several kinds with speckles or blotches of darker color, usually deep blue-purple, have also been found in sheltered dry sites and probably were grown in early Picuris.

Table 2.1 shows that eight-rowed and ten-rowed corn was common from 1150 to 1696, but when Spanish influence increased there was a corresponding increase in ears with many rows. Eight- to ten-rowed ears are rare today in the eastern Pueblos where Spanish influence was strong and are far more common in the Hopi villages and in Zuni. It will be interesting to see, when more Pueblo corn from 1690 to the present is excavated, how rapidly this change occurred and whether there were any periods of exceptionally sudden change.

When we study entire lots of corn from a site, from an entire level or time period of a site, or from a group of living Indians, we can expect peculiar distribution patterns for the number of ears with each row number. Thus, the bimodal curve of row numbers of the thirty-nine recent ears from Picuris, with peaks at twelve and sixteen rows, results from the combination of the white corn mode at sixteen and the blue corn mode at twelve rows (Table 2.2). A parallel can be seen in corn from the Santa Fe phase, A.D. 1225–1300 (Figure 2.2). The mode for the collection excavated in Area II is at eight rows, for Area VI at ten rows, and for Test Pit N at twelve rows. The Area II corn was composed of a rather uniform eight-rowed variety, while the other lots had few eight-rowed cobs. Clustered points on the graphs of row number and cupule width of cobs from various levels (Figures 2.1–2.5) show that throughout the period represented by these collections several kinds of corn were selected and grown. Unfortunately, since contact with Europeans, Indian corn has changed as a result of (1) less emphasis on traditional ceremony and consequent relaxation of rigid standards for the selection of seed corn, (2) less danger from crop failure or loss of seed stocks leading to less emphasis on hard, insect-resistant flints, and (3) increased exchange of corn with other peoples, including settlers from Mexico and the eastern states. The mixed character of modern corn from Picuris is evident from the wide dispersal of points on the graph of thirty-nine ears of the three most common kinds (Figure 2.6).

Table 2.1. Percent of Cobs of Each Row Number Found in Selected Southwestern Sites

Site and Date	Total Cob No.	Percent of cob-rows of each number				
		8	10	12	14	16+
Mogollon Area Sites						
Tularosa Cave, N.M., Georgetown Phase, A.D. 500–700	119	35	19	29	13	5
O-Block Cave, N.M., Three Circle Phases, A.D. 900–1000	136	52	29	12	6	1
O-Block Cave, Reserve Phase, A.D. 1000–1100	59	46	27	17	8	2
Hinkle Park Cliff Dwelling, Pueblo, N.M., A.D. 1100–1200	522	69	29	2	—	—
Casper Cliff Ruin, N.M., ca. A.D. 1300	42	43	40	17	—	—
Hooper Ranch Pueblo, A.D. 1200–1375	45	33	40	27	—	—
Carter Ranch Pueblo, A.D. 1450	349	23	41	31	4	1
Salado Area Sites						
Tonto Ruin, Room 16, A.D. 1400	1502	23	53	20	3	1
Anasazi Area Sites						
Basketmaker II and III, A.D. 500–700						
Mummy Cave, Canyon del Muertos, AZ	222	1	23	47	21	8
Mesa Verde, Step House, Pit Structure	322	9	22	46	18	5
MNA 7523, near Navajo Mt., AZ	697	11	14	47	18	10
MNA 2520, Turkey Cave, Segi Canyon, Arizona, Test I, Level II	159	21	21	47	9	2
Pueblo I and II, A.D. 700–1075						
Mesa Verde Site 1676	27	7	11	30	37	15
Antelope Cave, northern Arizona	1022	12	34	37	14	3

continued

Table 2.1. *Continued*

Site and Date	Total Cob No.	Percent of cob-rows of each number				
		8	10	12	14	16+
Pueblo I and II, A.D. 700–1075						
MA 2520, Turkey Cave, Tests 1 & 2 Levels 4–7	74	22	31	39	5	3
Kiet Siel, Arizona	133	30	36	32	2	—
Fremont Culture						
42 Un 95, Caldwell Village	125	6	18	59	11	6
Pueblo III						
42 Ka 433, Benchmark Cave, Glen Canyon 61.5 miles upstream from Lee's Ferry (before A.D. 1130)	90	30	36	29	4	1
42 Ka 274, Talus Rain (MNA 5369), Glen Canyon, 59 miles upstream from Lee's Ferry (A.D. 1160–1250)	507	23	35	36	5	1
42 Sa 583 Echo Cave, upper Glen Canyon, Utah	897	17	30	44	7	2
Mesa Verde, Step House (median of samples)	6024	30	34	28	1	—
Mesa Verde, Long House (median of samples)	2655	43	35	21	1	—
Las Madres (LA25) Galisteo, N.M., A.D. 1275–1350	135	7	27	49	13	4
Pueblo Largo, (LA183) Galisteo, N.M., A.D. 1275–1350	175	8	20	58	11	3
Picuris Pueblo						
Taos Phase, A.D. 1150–1225	14	29	35	29	7	—
Santa Fe Phase, A.D. 1225–1300	21	38	19	33	5	5
Vadito Phase, A.D. 1375–1490	23	22	43	35	—	—
San Lazaro Phase, A.D. 1490–1600	3	—	33	34	—	33

continued

Table 2.1. Continued

Site and Date	Total Cob No.	Percent of cob-rows of each number				
		8	10	12	14	16+
Picuris Pueblo, continued						
Trampas Phase, A.D. 1600–1696	324	18	28	44	7	3
Penasco Phase, A.D. 1930–present	40	—	8	32	15	45
Modern Pueblos						
Taos Pueblo						
White flour corn, A.D. 1953	18	—	—	22	33	45
Isleta Pueblo						
Blue flour-flint, A.D. 1953	8	—	—	12	25	63
Laguna and Mesita Pueblos (Keresan)						
Blue and blue-mixed ears, A.D. 1953	20	—	5	45	5	45
Moencopi (Hopi)						
Blue flint-flour, 1953	7	—	—	29	57	44

Table 2.2. Thirty-nine Ears of Corn from Modern Picuris (1953 and 1963)

	Percent of cobs of each row number				
	8	10	12	14	16+
White flour-dent	—	7	21	—	72
Blue flour-flint	—	12	41	18	29
Yellow flint	—	—	33	33	33

For comparison with the pre-Spanish Picuris corn we can use some collections from Moencopi, one of the Hopi villages, and from the Mohave Indians (Figure 2.7). Although there is far less Spanish influence in the Hopi villages than in those to the east, some is evident and the most useful comparison is with the white Mohave corn (Figure 2.7). This belongs to the variable Pima-Papago corn race (Anderson and Cutler 1942) and is one of the variants sometimes called Harinoso de Ocho (Wellhausen et al. 1952) or Eight-rowed Maize (Galinat and Gunnerson 1961). The common corn of most Pueblo III sites (Cutler and Meyer

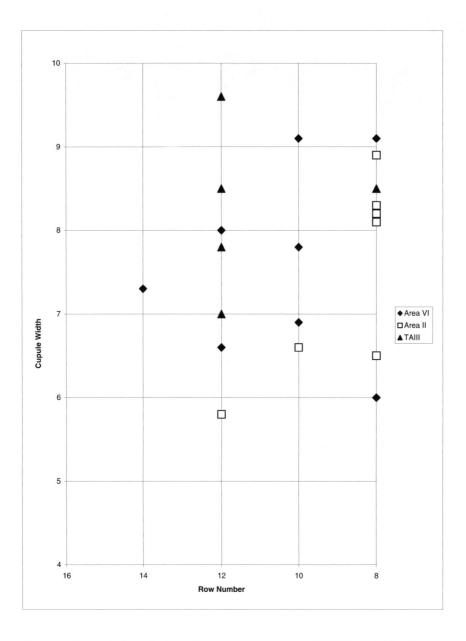

Figure 2.2. Corn cobs from the Santa Fe phase, A.D. 1225–1300. TAIII, Test Pit N, Levels 2, 3, 4, and 9 = ▲; Area II, Features 9 and 32 (Levels 6 and 8) = ☐; Area VI, Features 109, 116 (Level 2), 116 (Level 4), 124, and 127 = ◆.

Cultivated Plants from Picuris

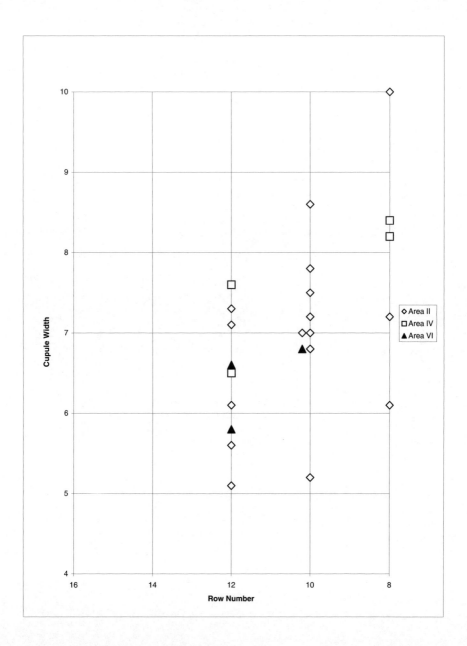

Figure 2.3. Corn cobs from the Vadito phase, A.D. 1375–1490. TAIII, Area II = ◇; Area IV = □; Area VI = ▲.

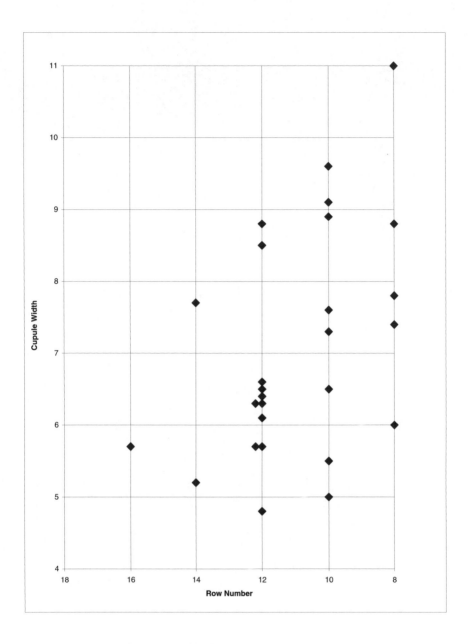

Figure 2.4. Corn cobs from the Trampas phase, A.D. 1600–1696. TAIII, Area VI, Feature 134, Level 3 floor = ♦.

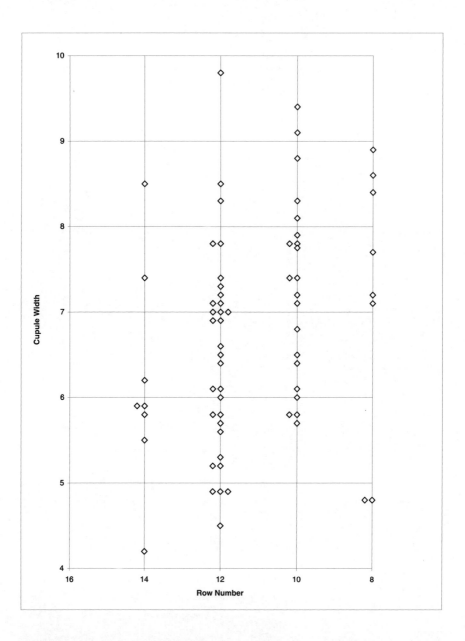

Figure 2.5. Corn cobs from the Trampas phase, A.D. 1600–1696. TAIII, Area II = ◇.

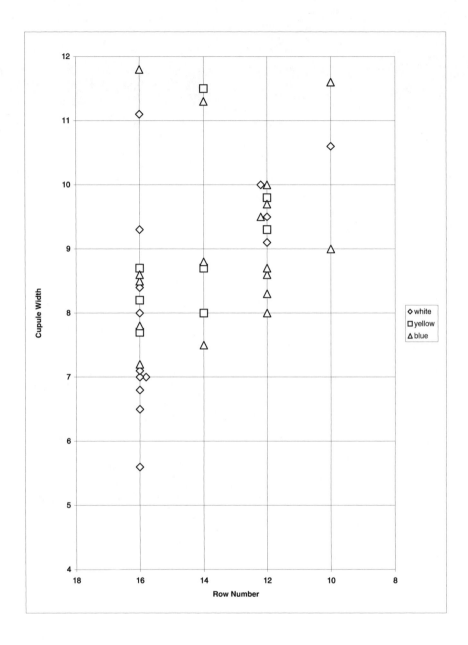

Figure 2.6. Corn from Picuris in 1953 and 1963. White = ◇; yellow = □; blue = △.

Cultivated Plants from Picuris

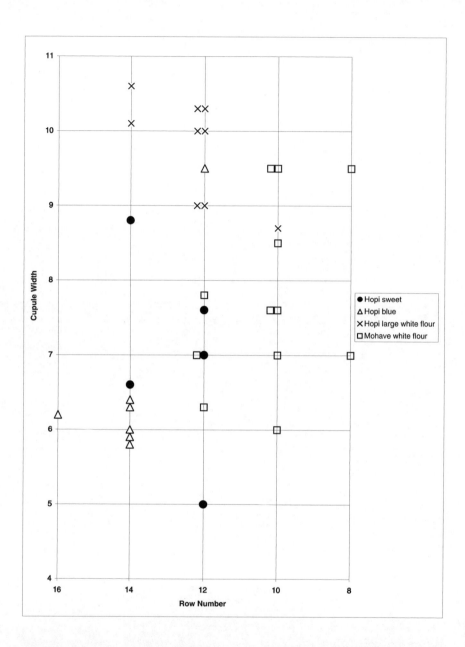

Figure 2.7. Maize from two reservations, 1953. Corn from a Hopi Pueblo, Moencopi: sweet = ●; blue = △; large white flour = ✕. Corn from Parker, Arizona, Mohave Indian Reservation: white flour = ☐.

34   *Cultivated Plants from Picuris*

1965; Cutler 1966) is the yellow flint form of this race and this most likely was the major corn grown at Picuris before 1690.

Squashes, Pumpkins, and Gourds

The oldest squashes and pumpkins in the Southwest are *Cucurbita pepo* (Cutler and Whitaker 1961; familiar modern varieties are the Halloween pumpkin, acorn squash, white bush scallop, and summer squash). About A.D. 700 *Cucurbita moschata* (modern forms are Kentucky field pumpkin and butternut squash) and *Cucurbita mixta*[1] (green-striped cushaw, the most common modern variety, is similar to most ancient kinds in the Southwest and is still grown in the Pueblos) reached the Southwest but, of the two, only *mixta* reached the Pueblo area. A squash from the Andes and west coast of South America, *Cucurbita maxima*, was brought to the Southwest by Europeans. The Pueblos grew, and still grow, varieties like hubbard and banana.

Cucurbita pepo

The oldest squash specimen from Picuris is a small, immature fruitstem (peduncle) of *C. pepo* from Test Trench 0, Level 9, a Taos phase (A.D. 1150–1225) level. Another carbonized, small (17 mm diameter at the base) peduncle came from the Vadito phase (A.D. 1375–1490) of Kiva B in Section I. About 400 variable, well-preserved, and uncarbonized seeds of *C. pepo* are from Feature 21 of Area V, which includes materials from Vadito through Trampas (A.D. 1600–1696) phases. These seeds have slightly thickened bodies and are similar to the dominant prehistoric variety found in northern New Mexico and Arizona. Practically all of the *C. pepo* seeds from Pueblo III sites in Mesa Verde are of this kind.

Cucurbita mixta

A small seed of the Taos cultivar of *C. mixta*, measuring 14.5 by 8 millimeters, was found in Area VI, Feature 7, Level 4, with Vadito pottery (A.D. 1375–1490, but probably occupied earlier). Similar seeds are occasionally found in Pueblo III sites and the same variety was grown recently in Taos Pueblo. A narrow seed (19.5 × 8.5 mm) from Area V, Feature 1, is similar to seeds of *C. mixta* from Oaxaca, Puebla, and Jalisco, Mexico. It differs from the usual prehistoric seeds and probably was introduced by the Spaniards. Three seeds of the green-striped cushaw type came from Area V, Feature 21 (Vaditos through Trampas phases). This variety of squash and hard-shelled strains of *C. pepo* are frequently found as containers in northern Arizona and New Mexico Pueblo III sites, and are sometimes used this way by the Hopi. Apparently they are used because the bottle gourd does not grow well in the region. It is likely that hard-shelled squashes were used as gourd substitutes at Picuris.

Cucurbita maxima

The only *C. maxima* specimens are eight large, brown-bodied seeds of the banana type found in Area V, Feature 21 (Vadito through Trampas). Several kinds of *C. maxima* have been grown by the Pueblo and other Indians of New Mexico and Arizona. I have seen specimens from a Navajo site in Canyon de Chelly, dated at about A.D. 1880, from Walapai, dated at about A.D. 1900, and from modern Pueblos.

Peach

Three small and broad peach stones marked with deep, rounded pits were found, two in Area V, Feature 5, and one from Test Pit B, Area III, Level 1, all running up to Spanish times. This kind of peach was introduced by the Spaniards and is still grown widely, although larger-fruited but probably less drought-resistant types were introduced later.

Beans

Common, or kidney beans (*Phaseolus vulgaris*), came from two areas: a few fragments from TAIII, Area II, Feature 34, and one and a half beans (size about 12 mm long and 6.5 mm wide) from TAI, Unit 3, Room 1 floor. All were carbonized.

1. The name of this plant is now *C. argyrosperma*. It is now known that *C. argyrosperma* ssp. *argyrosperma* reached the Midwest sometime shortly before A.D. 1000, and its presence at the Cahokia Mounds site is documented by recovery of an uncarbonized peduncle and by a reproduction of the fruit on a stone statuette (Fritz 1994).

3   Corn in the Province of Aminoya

Leonard W. Blake

(Written in 1974)

Leonard Blake's Comments, 1999

In St. Louis there is an amateur society, the Mound City Archaeological Society, a chapter of the Missouri Archaeological Society, which was organized in the early 1960s. In the 1970s, some of the members used to write short articles about their archaeological interests that were printed in a newsletter, which was circulated among the members and a few others. This paper was written for that purpose and printed in February 1974 in the *Newsletter of the Mound City Archaeological Society*. It was reprinted in the April 1974 *Quarterly Newsletter of the Illinois Association for the Advancement of Archaeology*. The IAAA is a group of Illinois amateur archaeologists, organized a few years earlier, which then still had limited membership. We are assured by its secretary that it may be again reprinted here, for publications of the IAAA, purposely, are not copyrighted.

I wrote this paper for fun, as a parody of some archaeological articles, and, it must be confessed, as a result of a rather childlike delight in playing with figures. After completion, it was realized that second-hand testimony of a romantic writer, who was not a part of the De Soto expedition, could not be taken seriously. Also, no one knows whether Garcilaso de la Vega is talking about "corn-on-the-cob" or about "shelled" corn, that is, corn kernels removed from the cob. There is archaeological evidence that some protohistoric Indians in northern Arkansas stored their corn on the cob in above-ground structures (Blake and Cutler 1979). Weights per bushel used in this report are those of shelled corn. Calling my paper a "parody" gives an excuse, of sorts, for its production. No one archaeological report is targeted, but it should not be too hard to find one.

When the members of the De Soto expedition again saw the Mississippi after their nearly disastrous attempt to reach Mexico overland, according to Garcilaso, they came upon " . . . two towns, one near the other, and each comprising 200 houses . . . On entering these places the Castilians found a great quantity of corn, and other grains and vegetables as well as such dried fruits as nuts, raisins, prunes, acorns, and some additional ones unknown to

Spain ... Alonso de Carmona declares that on measuring the corn found in both settlements, they discovered that there was by count eighteen thousand bushels of it ..." (Varner and Varner 1951:531–32).

Castetter and Bell (1942:52) indicate that the corn yields of the Pima seldom exceed 10 to 12 bushels per acre. Will and Hyde (1968:142) said that yields of 20 to 25 bushels an acre were considered unusually good, while agriculture was practiced using Indian methods on the Upper Missouri.

If a figure of 25 bushels per acre is used, 18,000 bushels represents the production of at least 720 acres, or corn fields totaling more than a square mile. If an estimate of 12 bushels per acre is used, 1,500 acres or nearly 2 and $\frac{3}{8}$ square miles of corn fields would be required to produce 18,000 bushels. The latter figure seems more probable, despite the fertility of the alluvial valley of the Mississippi, because Indian cultivators often grew other crops such as squash and beans, mixed with the corn.

Will and Hyde (1968:108) figure a family as six persons in their calculations of the number of acres cultivated per family while native agriculture was still in its primitive state on the Upper Missouri. According to their information, the area in cultivation of all crops came to $\frac{1}{3}$ to 1 acre per person, or 2 to 6 acres per family.

Using six persons per house gives an estimated population of 2,400 persons for the two towns with a total of 400 houses. Seven hundred and twenty acres in corn seems a bit on the low side, as it amounts to less than $\frac{1}{3}$ of an acre per person. One thousand five hundred acres is equivalent to $\frac{5}{8}$ of an acre per person, when population is estimated at 2,400. If one calculates five persons per house, the estimated population is 2,000. With this population, average acreage per person is less than $\frac{3}{8}$ of an acre, if yields are 25 bushels; or $\frac{3}{4}$ of an acre, if yields are 12 bushels. The latter population estimate produces a per capita acreage use closer to Will and Hyde's data for the more sedentary tribes.

If the population was 2,400 people, 18,000 bushels would provide 7.5 bushels, or 410 pounds, of corn for each man, woman and child, as corn weighs about 56 pounds per bushel. When corn is eaten, it produces approximately 1,640 calories per pound. This means, if Garcilaso is not exaggerating and if our estimates are not too far out of line, that corn alone produced close to 1,840 calories per day, per capita (1,640 × 410/365) exclusive of those from fish, game, other cultivated plants, nuts, fruits, etc. If the population was 2,000 instead of 2,400, 18,000 bushels would provide each person with 9 bushels or 504 pounds of corn, equivalent to 2,264 calories per day.

Alas, these calculations may be meaningless. It must be remembered that Garcilaso was a chronicler, not a member of the expedition. His information was all second hand, quite old, and subject to the inaccuracies of time and distance. The account of the Gentleman of Elvas, who was actually present, mentions the two towns in the province of Aminoya, but says nothing about the number of houses in each. He does, however, estimate the amount of maize found in the towns. "The Christians chose for their quarters, what appeared to be the best town . . . The maize that lay in the other town was brought there, and when together the quantity was estimated to be six thousand fanegas" (Bourne 1904:vol. 1, p. 186). A "fanega" is approximately equivalent to our bushel. But wait! Even 6,000 bushels is a lot of corn: 336,000 pounds or 168 tons . . .

4 Corn from Three North Carolina Sites, 31Gs55, 56, and 30

Leonard W. Blake
Washington University

(Written in 1987)

Leonard Blake's Comments, 1999
    The three North Carolina sites in this report were excavated under the direction of Dr. Janet Levy of the University of North Carolina at Charlotte and Dr. Alan May of the Shiele Museum of Natural History in Gastonia, North Carolina. Estimates of dates and determination of cultures are those of the excavators.

    Samples of carbonized corn cobs and cob fragments were received from three sites in Gaston County, North Carolina, from Dr. Janet Levy of the University of North Carolina at Charlotte. In this report corn cob samples from each site are considered separately, compared with the others, and compared with corn from an early historic site in Virginia and from a later Spanish Mission off the coast of Georgia.

    Samples were received from three different locations at 31Gs55, which is described as a multicomponent site with occupations ranging from Middle Archaic into Late Prehistoric, which is probably "South Appalachian Mississippian." There is a carbon 14 date of A.D. 1600 ± 50 on wood from Feature 40 and another from a different part of the site of A.D. 1350 ± 70 (Letter from Dr. Janet Levy, Jan. 12, 1987).

    Although mean row numbers vary between the rather small samples from Trench D and from Feature 39, when combined they average the same as corn from Feature 40 with a mean row number of 8.9. When row numbers and cupule widths of the individual specimens measured are placed on a coordinate graph, they cover the same area, with only a few exceptions (Figure 4.1). I consider them to be very similar examples of the race of corn that was called Northern Flint by Brown and Anderson (1947) but, more recently, Eastern Eight Row by Cutler and Blake (1976). The later name seems preferable to me because the race is predominantly eight-rowed and the kernels may be flint, flour, or sweet. Although it was dominant prehistorically in the northern United States,

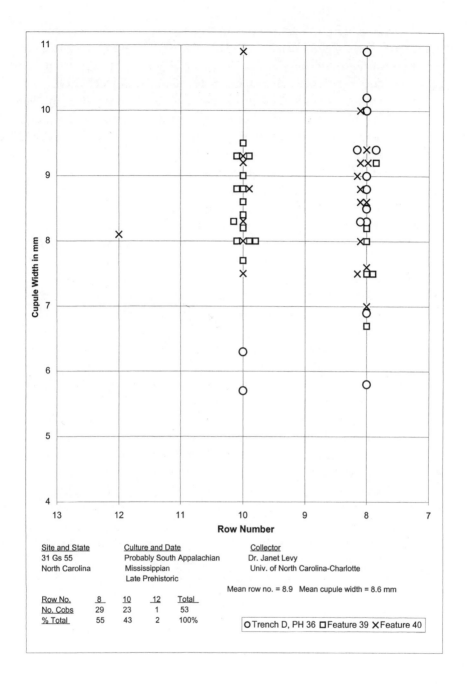

Figure 4.1. Corn from 31Gs55, North Carolina.

it was also present by A.D. 1000 in the southeastern states, where average row numbers are often slightly higher than in the Northeast (Blake 1986:3). Because the samples from 31Gs55 appear to be comparable, the A.D. 1600 ± 50 date of Feature 40 does not appear to be out of line for all three samples.

A sample of corn cobs was received from only one location at 31Gs56, that is, from Test Pit 2, Feature 1, described as a large burned area of corn and wood. I was able to obtain measurements on sixteen cobs (Figure 4.2). The distribution of row numbers is not the same as that of the combined locations of 31Gs55, but the mean row numbers are identical. While the median size of the cob cupules from 31Gs56 is 8.1 millimeters, somewhat less than the 8.6 millimeters of 31Gs55, the cobs are similar and are considered to be of the same race.

Samples were received from two locations at 31Gs30, which is described as representing a Late Prehistoric or, possibly, Historic occupation of about A.D. 1500–1600 (Levy, letter, Jan. 12, 1987). Each measurable sample is small, consisting of ten cobs from Trench B and nine from Trench F, Feature 24 (Figure 4.3). The combined mean row number on the total of nineteen cobs of 8.8 nearly equals the 8.9 of the other sites. The cobs are very much smaller, however, median cupule width being only 5.5 millimeters, 64 percent of the 8.6 millimeters of 31Gs55 and 68 percent of the 8.1 millimeters of 31Gs56. The kernels are also thinner, the median being only 2.9 millimeters as against 3.4 millimeters for the corn from the other two sites (see Definition of Terms Used, below, for measurement of kernel thickness on corn cobs).

If the smaller size was the result of a bad crop year, or neglect, a decline in the mean row number might be expected, for under adverse growing conditions the row number of corn ears tends to decline (Emerson and Smith 1950:7). Because the mean row number is nearly equal to that of the larger cobs from the other sites, the small size does not appear to be the result of stress. John Banister, writing in April 1679 in Virginia, said: "The Indians have two Sorts more of Rath-ripe Corn [that is, early corn], the ears of ye lesser Sort are no bigger than ye haft of a knife, and its stalks not much higher than ones middle, of these they can make two Crops in a Year" (Ewan and Ewan 1970:40–41). It is suggested that some, if not all, of the small cobs from 31Gs30 may be the early corn described by John Banister.

Corn from 44Sn22, a site on tidewater in Virginia, possibly of Nottaway Indians of about A.D. 1500–1620, and from 9Lb8, a historic Spanish Mission of about A.D. 1590–1670 on an island off the coast of Georgia, is compared with corn from 31Gs55, 56, and 30 in Table 4.1. Mean row numbers and distribution of the proportions of eight- and ten-rowed cobs are quite similar. Median cupule widths of the corn cobs from both of these sites are somewhat less than those from 31Gs55 and 56, but greater than those from 31Gs30. Corn from all of these sites appears to be the southern variant of the Eastern Eight Row race.

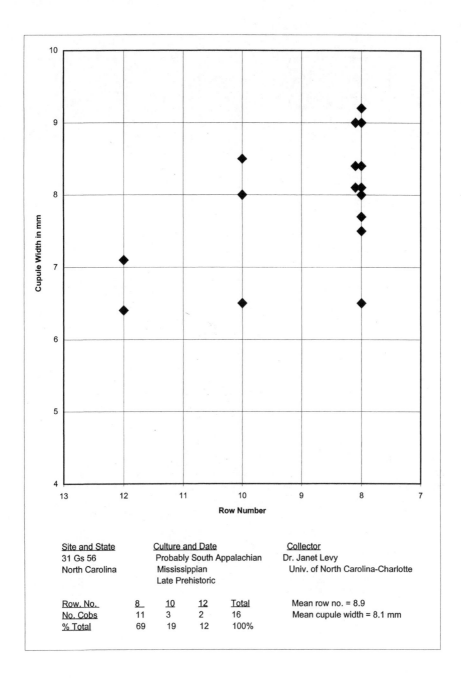

Figure 4.2. Corn from 31Gs56, North Carolina.

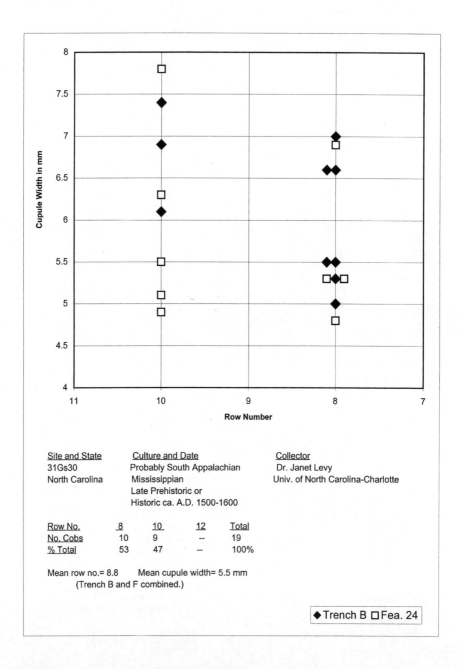

Figure 4.3. Corn from 31Gs30, North Carolina.

Table 4.1. Comparative Samples of Corn from North Carolina, Virginia, and Georgia

Site	Date, Culture, Tribe	Mean Row No.	Median Cupule Width in mm	No. Cobs	Row Numbers % of Total Cobs				
					8	10	12	14	16+
North Carolina									
31Gs55	Late Prehistoric probably South Appalachian Mississippian	8.9	8.6	53		55	43	2	
31Gs56	Approximately the same as 31Gs55	8.9	8.1	16		69	19	12	
31Gs30	Late Prehistoric or Historic, ca. A.D. 1500–1600	8.8	5.5	19		53	47		
Virginia									
44Sn22	Perhaps Nottaway Indian, ca. A.D. 1500–1620	8.9	7.3	110		60	35	5	
Georgia									
9Lb8	Historic Spanish Mission, A.D. 1590–1670	8.8	7.5	20		60	40		

Definitions of Terms Used

Cupule Width: "The distance across the entire pocket in which a pair of grains (kernels) and their spikelets is borne is a measure of the size of the cob" (Cutler and Blake 1976:4).

Kernel Thickness: "The distance along the cob axis from the base of one set of lower glumes to the upper edge of their cupule is exactly equivalent to the thickness of each kernel produced there" (Nickerson 1953:83). It has been our practice to measure five cupules along the cob axis and to divide this measurement by five to obtain a more accurate figure.

5 Cultivated Plant Remains from Historic Missouri and Osage Indian Sites

Leonard W. Blake

Washington University

(Written in 1986)

Leonard Blake's Comments, 1999

This paper was one of several given by different people in a symposium on the Missouri and Osage tribes, organized by Dr. Carl H. Chapman, given on April 25, 1986, at the Annual Meeting of the Society for American Archaeology at New Orleans.

We had been receiving plant remains from excavations at Missouri and Osage sites by the University of Missouri-Columbia for a period of over 20 years.

Cultivated plant remains from two Missouri, two Little Osage, and two Big Osage sites are discussed here. They consist of carbonized specimens picked from deposits and those recovered by flotation. Others have reported on the amount of excavation and recovery at the various sites. I have been asked to report separately on three parts of the Utz site, occupied by the Missouri Indians, that is, 23Sa2, 23Sa2B, and 23Sa2C. Recovery of trade items at the last indicates historic, post-contact occupation at that part of the site.

Table 5.1 gives comparative figures on the carbonized corn cobs and cob fragments recovered. All but a very small part are of a race called Eastern Eight Row (Cutler and Blake 1976), formerly Northern Flint (Brown and Anderson 1947), which usually has eight, sometimes ten, and rarely twelve rows of kernels. This race of corn reached as far north and east as Ontario, Canada, by about A.D. 800 (Stothers 1976:156). By A.D. 1200 it dominated most of the country east of the Mississippi, and, by A.D. 1500, most of the region east of the Rockies (Cutler and Blake 1976:5).

Corn from the sites considered here has been described by Dr. Hugh C. Cutler as follows: "The mean row number is slightly higher [than that of most examples of this race], the grains a bit thinner and the cupule not quite as elevated, differences pointing to varieties of areas farther south" (Cutler n.d.).

Table 5.1. Comparison of Corn Cobs from Historic Missouri and Osage Sites and from a Historic Kickapoo Site

	No. Cobs	Mean Row No.	Median Cupule Width (mm)*	% of Total Cobs Row Number				
				8	10	12	14	16+
Missouri								
Gumbo Point								
(23Sa4) 1712–1794				No measureable cobs or fragments were recovered				
Utz								
(23Sa2C) 1650–1712	47	9.2	8.8	51	41	6	2	—
(23Sa2) ca. 1460–1712	33	9.2	8.6	51	40	9	—	—
(23Sa2B) ca. 1460–1712	15	9.2	7.4	60	20	20	—	—
Little Osage								
Coal Pit								
(23Ve4) 1790–1820	459	8.8	9	65	29	5	1	—
Plattner								
(23Sa3) 1714–1794	6	10.3	6.5	33	17	50	—	—
Big Osage								
Carrington								
(23Ve1) 1775–1820	76	8.9	8.3	66	22	11	1	—
Brown								
(23Ve3) ca. 1650–1777	87	9.1	8.7	55	38	6	1	—
Corn cobs from another site for comparison								
Kickapoo								
Rhoads								
(11Lo8) 1760–1820	262	8.4	8.3	85	12	3	—	—

All corn cobs are carbonized.
*Cupule width, the distance across the entire pocket in which a pair of grains and their spikelets is borne, is a measure of the size of the cob. (Cutler and Blake 1976:4).

Table 5.1 illustrates a typical example of more northern Eastern Eight Row from the Rhoads Kickapoo site of A.D. 1760–1820 in Illinois.

Several of the small twelve- and fourteen-rowed cobs from Utz and from most of the Osage sites appear to be of a race of popcorn called North American Pop (Cutler and Blake 1976), an early race grown by some of the Plains Indians (Will and Hyde 1968:305, 307), which has persisted into the present. At the relatively late Coal Pit site several very large twelve- and fourteen-rowed cobs may be the result of slight mixing of the large, many-rowed Southern Dent corns being grown by white farmers in the South in the eighteenth and nineteenth centuries (Brown and Anderson 1948; Anderson and Brown 1952), but it was not sufficient to alter the overall composition of the corn.

It will be noted that the mean row number of corn from the later Osage sites of Coal Pit and Carrington, which date into the nineteenth century, is slightly lower and the proportion of eight-rowed corn greater than that of corn from Utz and Brown, which were abandoned in 1712 and 1772, respectively. This could be due to increased influence of corn from the north and east acquired from Indians or whites.

Part of the large collection of corn cobs from Coal Pit was recovered from "smudge-pits" used to smoke deer hides for trade (Swanton 1946:443–47; Binford 1967).

Historic Indians on the Plains and in the eastern United States had and some still have a number of varieties of beans distinguished largely, but not entirely, by color (Wilson 1977:84; Weltfish 1971:147; Waugh 1916:103–8, plate 34). Carbonized beans (*Phaseolus vulgaris*) were recovered from all of the sites considered here except the late Missouri Gumbo Point site. Carbonization destroys the color and often the seed coat, as well, so that halves of beans with smaller dimensions remain. In the case of the rather large collection of beans from the historic 23Sa2C part of the Utz site, we have presented separate and combined figures on whole and half beans in Table 5.2.

Beans from the 23Sa2C part of the site appear to resemble those from the early contact King Hill site in St. Joseph, Missouri, more than collections from the other more distant sites shown for comparative purposes in Table 5.2.

Beans were recovered from all four of the Osage sites, but not in sufficient quantity for significant comparisons. Considering the relative scarcity of carbonized beans at Utz, except at the historic 23Sa2C part, one wonders "Who spilled the beans?" into the fire. It appears more than coincidental that there are so many from the historic part of the site and so few from elsewhere.

Squash (*Cucurbita pepo*) seeds were recovered from all of the sites considered here, except Gumbo Point, one part of Utz (23Sa2B), and Plattner. Most were preserved uncarbonized by copper salts from contact with objects of trade brass or copper. It will be noted from Figure 5.1 that a number of the seeds are

Table 5.2. Beans (*Phaseolus vulgaris*) from Historic Missouri and Osage Sites and Three Other Historic Sites

	No. of Beans*	No. of Measurable Beans	Range (mm)		Median (mm)		
			Length	Width	Length	Width	L/W
Missouri							
Gumbo Point (23Sa4) 1712–1749	None						
Utz (23Sa2C) ca. 1650–1712							
Whole beans	—	42	14.9–9.0	7.8–4.6	11.4	6.3	1.81
Half beans	114	89	15.2–8.2	7.9–3.6	11.1	6.0	1.85
All beans	156	131	15.2–8.2	7.9–3.6	11.2	6.2	1.81
(23Sa2) ca. 1460–1712	18	16	13.7–8.0	7.2–5.2	12.2	6.3	1.94
(23Sa2B) ca. 1460–1712	12	9	12.7–8.5	8.0–4.5	11.4	6.2	1.84
Little Osage							
Coal Pit (23Ve4) 1790–1820	11	5	9.9–5.9	6.3–3.6	9.4	4.7	2.00
Plattner (23Sa3) 1714–1794	4	2	9.5–9.0	6.0–5.2	9.3	5.6	1.66
Big Osage							
Carrington (23Ve1) 1775–1820	Fragments Only						
Brown (23Ve3) ca. 1650–1777	3	3	11.2–8.8	6.3–5.0	10.8	5.5	1.96

continued

Table 5.2. Continued

	No. of Beans*	No. of Measurable Beans	Range (mm)		Median (mm)		L/W
			Length	Width	Length	Width	
Beans from other sites for comparison							
Kickapoo							
Rhoads							
(11Lo8) 1760–1820		41	16.7–7.0	8.7–3.8	9.4	5.7	1.65
Late Oneota (Kansa?)							
King Hill							
(23Bn1) ca. 1700		31	15.8–8.0	7.5–4.4	11.0	5.9	1.86
Kaskaskia							
Zimmerman							
(11Ls13) 1673–1691		38	12.3–7.2	7.6–3.3	9.5	5.0	1.90

All beans are carbonized.
* Whole and half beans, plus uncounted fragments.

unusually narrow, some even having a length more than twice the width. At the suggestion of botanist Dr. Charles Heiser of Indiana University, a copy of this chart was sent to Deena Decker, a Ph.D. candidate at Texas A. and M. University whose dissertation research focuses on systematics of *Cucurbita* species. In a letter dated December 13, 1985, she replied in part: "I noticed that there is quite a bit of diversity in the size and shape of the Missouri and Osage seeds, covering the range of variation with which I am familiar. The smallest seeds probably represent ornamental gourds (*C. pepo* var. *ovifera* and/or *C. texana*). At the other extreme are the long and narrow seeds mentioned. These also appeared at the Florida site [on which she had reported] after about A.D. 1700. I believe that they represent a cultivar, or landrace introduced from Mexico." She went on to say that she had collected these relatively narrow seeds from characteristically ridged fruits in Veracruz, Oaxaca, and in four other locations in Mexico.

Tiny fragments of squash rinds were recovered from all of the sites by flotation. They were most abundant at Plattner where a few fragments of gourd rind (*Lagenaria siceraria*) were also present. It is not known with certainty whether all of the squash rinds are from the cultivated *Cucurbita pepo* squash or from one of the wild forms. It seems reasonable to expect that some, if not all, are the cultivated form.

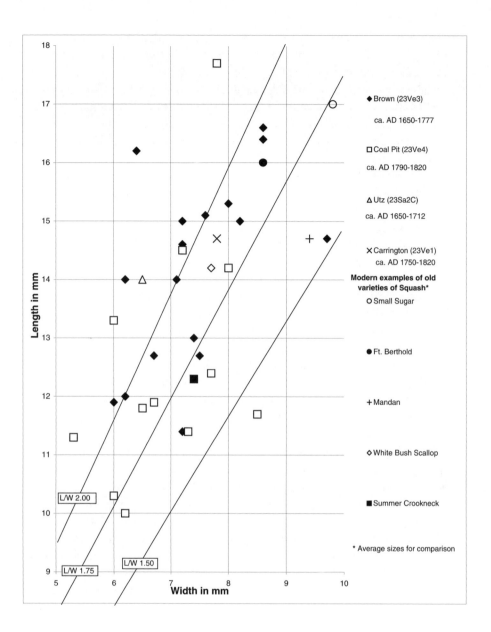

Figure 5.1. Measurable squash seeds from Historic Missouri and Osage sites.

Seeds of watermelon (*Citrullus lanatus*), a plant that originated in the Old World (Whitaker and Davis 1962:2), were found at several locations at both the 23Sa2C part of Utz and at Coal Pit. Part of the inside of a carbonized seed from the Brown site may be watermelon, but we were unable to identify it with certainty. Measurable seeds from Utz and all but one from Coal Pit range in size between 9.9 by 6.0 millimeters and 10.3 by 6.8 millimeters. One from Coal Pit is somewhat larger, 11.5 by 6.0 millimeters. It may represent a different variety. All are carbonized.

The watermelon was brought to Florida by the Spanish before 1576 (Connor 1925:159) and within about 100 years was grown by Indians from Mexico to southern Canada and from the east coast to California (Blake 1981:194). Reasons for rapid acceptance include the refreshing sweetness of the watery flesh and the ease of growing the plant in the hot, continental climate of the present United States, using the same familiar methods by which squashes are grown.

In August 1806, Pike's Lieutenant Wilkinson was given watermelon while visiting the Little Osage village (Jackson 1966:vol. 2, p. 5), which Chapman (1974:120) indicates was probably the Coal Pit site.

Carbonized stones of peaches (*Prunus persica*) were recovered from the 23Sa2C portion of the Utz site and at the Coal Pit and Brown sites.

According to ethnobotanist Elizabeth Sheldon (1978), peaches were probably introduced into the United States by the Spanish before the eighteenth-century arrival of the English in the Carolinas and the French in Alabama and Louisiana. Peach remains from a well at St. Augustine, Florida, used between 1594 and 1623 provide archaeological evidence for an early introduction.

The pit of a single domestic cherry, probably Sweet Cherry (*Prunus avium*), was recovered from Feature 110 in excavations in the 1960s at Coal Pit. *Prunus avium* is an Old World plant introduced by European settlers.

The following summarizes evidence, which has been mentioned, on outside influences on the cultivated food plants of the Missouri and Osage:

> Corn (*Zea mays*)—There appears to be some slight influence of the Southern Dent corns, grown by white farmers, in the corn from Coal Pit, but not enough to alter the overall composition of the collection. Corn of both the latest sites, that is, Coal Pit and Carrington, indicates influence of the more extreme forms of the Eastern Eight Row race, which was grown by both Indians and white farmers to the north and east.
>
> Squash (*Cucurbita pepo*)—Some of the squash seeds from most of both early and late Missouri and Osage sites show evidence of influence by a Mexican variety of squash or pumpkin, which was also present in Florida after 1700.
>
> Watermelon (*Citrullus lanatus*)—the Old World plant.
> Peach (*Prunus persica*)—the Old World plant.
> Domestic Sweet Cherry (*Prunus avium*)—all Old World plants.

Watermelon and peach are present in collections from both the early historic 23Sa2C part of Utz, which was abandoned in 1712, and the late historic Coal Pit site where a single stone that appears to be that of the domestic sweet cherry was also found. Peach and possibly watermelon were found at the Brown site, which dates from about 1650 to 1777.

6 Corn for the Voyageurs

Leonard W. Blake
Washington University

(Written in 1994)

Leonard Blake's Comments, 1999

Managers of the tourist facilities at the partly reconstructed Fort Michilimackinac, in Mackinaw City, Michigan, were considering construction of drying racks for corn, such as were mentioned in some of the historic accounts, but no one knew what they looked like. Blake was asked to find out what he could about the processing of corn for use by the voyageurs. This paper is the result. Unfortunately, he was not able to find any precise specifications for the reconstruction of drying racks, like those used at the fort.

The race of corn (*Zea mays*) grown in the northeastern United States and southeastern Canada was called Northern Flint by Brown and Anderson (1947). More recently, it has been called Maiz de Ocho by Galinat (Upham et al. 1987) and Eastern Eight Row by Cutler and Blake (1976). The race was present in most of the eastern United States at the time of first European contact. The ears usually have eight rows of kernels. These are often flint, but sometimes may be flour or sweet corn. Flint kernels have soft starch in the center surrounded by a layer of hard, flinty starch.

Recent studies have demonstrated that this is a most unusual race of corn, which evolved rapidly (Doebley, Goodman, and Stuber 1986). It is early maturing (Brown and Anderson 1947) and drought and cold resistant (Will and Hyde 1968:73). These characteristics enabled the Indians to grow it as far north as North Dakota and the lower St. Lawrence River Valley (Yarnell 1964:128). The hard starch of the flint kernels makes the ears resistant to rot and insect damage, but causes some problems in efforts to convert the kernels into food for human beings.

Waugh (1916:79) quotes LeJeune, a seventeenth-century Jesuit observer, who said that he had seen twenty ways of cooking corn by the Hurons and added that many more had been reported by others. However, there appear to have been a limited number of basic ways of preparing the hard flint corn for use on journeys.

Henry Ellsworth (1937:36) described a method used by the Osage: "The corn is taken up before quite ripe and boiled—it is then dried and hung up until wanted. [After being shelled] it is finally pounded or ground and sifted—it will keep for a long time and no article can be packed for a long journey, containing so much sustenance in so little weight." Sometimes green corn or mature corn was parched rather than boiled and treated, in a similar manner with pounding and sifting, to make a lightweight, nourishing trail food. There is a good description of the method of processing mature corn in Heckwelder (1971:195).

Parker (1910:69), an authority on the Iroquois, has described another way of preparing corn for a long journey, more suitable for mass production. He reported that after flint corn was shelled "it was boiled for 15 to 30 minutes in a weak lye made of hardwood ashes." The result was a form of hominy. Research has demonstrated that the use of an alkali such as wood ash or lime decreases the total nutritional value of corn but increases its nutritional quality by increasing the ratio of essential amino acids (Katz, Hediger, and Valleroy 1974). It thereby avoids the dietary deficiencies of an exclusive diet of corn, not processed in this manner, which can lead to the development of diseases such as pellagra.

Parker (1910:69) continues,

> When the hulls and outer skins had been loosened looking white and swelled, the corn was put in a hulling basket then taken to a brook or large tub, where it was thoroughly rinsed to free the kernels of any trace of lye and to wash off the loosened hulls and skins. The granules were sifted through the meal sieve to make the meal fine and light. After this process, the meal was mixed with boiling water and quickly molded into a flattened cake. . . . The cake was then plunged into boiling water and cooked for nearly an hour. . . . sometimes the molded loaf is baked instead of boiled, especially for long journeys. The baked loaf, if not wet, will not become moldy like boiled bread, and it is the approved form for hunting and war parties.

White fur traders adopted a similar method of processing corn, but divided the procedure. Part was completed before setting out and part daily before consumption. Alexander Henry, a fur trader, wrote in 1761: "This species of corn [that is, Indian corn] was prepared for use by boiling it in strong lie, after which the husk may be easily removed; it is next mashed and dried. In this state it is soft and friable like rice. The allowance for each man on a voyage is a quart a day; and a bushel with two pounds of prepared fat, is reckoned to be a month's subsistence." He added that "no other allowance is made of any kind;

not even of salt" and that "the men are capable of performing their heavy labor" (Henry 1901:54; Armour 1978:33). Thirty-two years later, John Macdonell, while at the Grande Portage on western Lake Superior, wrote, "A full allowance to a voyageur while at this poste is a quart of Lyed Indian Corn or maize, and one ounce of Greece" (Gates 1933:950). Still the same ration was used.

A few years after this, the famous trader and explorer Alexander Mackenzie noted that the food of the voyageurs was still Indian corn and melted fat. He described the process of preparing the corn and added that after it was washed "it was carefully dried on stages." He also described how the prepared corn was cooked on the voyage. "One quart of this is boiled for two hours over a moderate fire, in a gallon of water; to which, after it has boiled a small time, are added two ounces of melted suet; this causes the corn to split, and in the time mentioned makes a pretty thick pudding. If to this is added a little salt, (but not before it is boiled, as it would interrupt the operation), it makes a wholesome, palatable food, and easy of digestion. This quantity is fully sufficient for a man's subsistence during twenty-four hours; though it is not sufficiently heartening to sustain the strength necessary for a state of active labor" (Mackenzie 1966:xlvi, xlvii). It will be noted that Mackenzie has given evidence that the prepared corn was dried on stages, and also that his men got two instead of one ounce of grease a day plus a little salt. There is a footnote to Mackenzie's account that reads as follows: "Corn is the cheapest provision that can be procured, though from the expense of transport, the bushel costs about twenty shillings sterling, at the Grande Portage. A man's daily allowance does not exceed ten-pence" (Mackenzie 1966:xlvii).

The narrative of trader David Thompson tells when and how the processed corn was cooked and served on his voyage to Sault Ste. Marie in 1798. He wrote:

> The provisions we had to live on were hulled corn, part of a bag of wild rice, with a few pounds of grease to assist the boiling. It was customary after supper, to boil corn or rice for the meals of next day, and in good weather we set off by 4 A.M., the Kettles were taken off the fire in a boiling state and placed in the Canoe, and two hours afterwards we had a warm breakfast; if Lightening and Thunder came in the day the Corn became sour and had to be thrown away; but the rice never soured; the same thing at night, when the kettle had corn it was soured, but if of rice it kept good: the Men assured me that the Lightening and Thunder had no effect on the wild rice; and that in the heats of Summer the Corn soured so frequently, they were half starved; to boil a Kettle of corn requires three to four hours. The rice is cooked in half an hour, but it is very weak food. All the Corn for these voyages has to be steeped in hot lye of wood ashes to take off the rind of the grain. On the 28th May we arrived, Thank God at the Falls of St. Maries. (Glover 1962:218)

There may be some question whether the souring was real or whether the crew was using it as an excuse for a change from the monotony of the corn diet.

In the 1776 inventory of the estate of the trader and post commissary John Askin (Armour and Widder 1978:209, 212, 218), there are items that appear to apply to the preparation of corn for the voyageurs: "Stages for drying corn worth 8 Pounds"; "2 large Kittles in a furnace at Water Side, 20 Pounds"; "2 Hulling Basketts @ 8 Shillings, 16 Shillings"; "$1\text{-}\frac{1}{2}$ Barrell of Ashes @30 Shillings, 2 Pounds, 5 Shillings."

Although it is reasonable to believe that the drying stages were used for drying prepared hominy corn, they may have also been used for drying unshelled ears of corn. Wilson (1977:46, 47), in his account of the Hidatsa Indians of North Dakota, shows that it was their regular practice to dry all corn on stages after it was picked. It appears that this was also done in the Upper Great Lakes region, for John Cornell, in a letter to John Askin dated October 18, 1788, wrote, "I will shell out my corn as soon as it is dry" (Quaife 1928:263). Also, Kinietz (1965:322) quotes the Memoir of Raudat (Letter 45, 46) that corn was regularly picked before it was fully ripe at such places as Sault Ste. Marie where fogs delayed the maturity of corn in the short growing season.

It may be of interest to learn where the corn came from that was used to provision the voyageurs. Lotbiniere said in 1749 that the Indians used to grow corn a few years earlier on the hills in back of Fort Michilimackinac, but that "4 or 5 years of production impoverishes it completely, and once it has reached that state it can not recondition itself" (Gerin-Lajoie 1976:6).

Some corn was apparently still grown at the fort in 1754, for in that year J.C.B., an unidentified French soldier, visited there and indicated that soldiers stationed there were cultivating some corn. He wrote, "The Fort is surrounded by a stockade, mounted with six cannon, and has thirty men in garrison who are changed every three years, if they wish. Their only remuneration is powder and lead bullets. This is enough, because they cultivate maize or Indian corn, and go hunting and fishing, thus supplying their needs" (Stevens, Kent, and Woods 1941:17).

John Askin's diary shows that he began to plant corn on May 23, 1774 (Quaife 1928:53), but, as Heldman (1983:12) points out, he does not mention whether it was at the fort or at his farm three miles south-southwest of present-day Mackinaw City (Quaife 1928:10). He was probably successful in growing corn over a period of years because of his practice of manuring and crop rotation (Quaife 1928:14).

In spite of this evidence that some corn was grown near the fort, it is certain that this was not enough to supply the needs of the traders to provision their employees for their long and arduous trips. One major source of supplies of corn was the settlement of L'Arbre Croche, now known as Cross Village, twenty miles south of the fort by land and somewhat more by water. The Ottawa, who formerly lived near the fort, moved there in 1742 when the French Jesuit missionaries built the mission church of St. Ignace in that place (Heldman 1983:8). John Askin, shortly after he arrived at the fort in 1764, acquired in partnership a large farm near the Mission.

Alexander Henry (1901:49) wrote: "The Otawas of L'Arbre Croche, who, when compared with the Chipeways, appear to be much advanced in civilization, grow maize, for the market of Michilimackinac, where this commodity is depended upon, for provisioning the canoes."

In 1714, a year before the fort was built on the south side of the straits, the Potawatomi were supplying corn to Mackinac from an island in Green Bay in western Lake Michigan (Kinietz 1965:313). Although the Indians at Green Bay later became best known as a source of wild rice, it is possible they may have also supplied some corn. Peter Pond remarked of the area that "the Inhabitans Rase fine Corn" (Gates 1933:33).

Also in 1714, 800 minots of corn were exported from Detroit. (A minot is equivalent to 1.10746 bushels, according to McDermott 1941:104.) It was urged then that the port, Detroit, be preserved and that some Indians remain there because the land was good "particularly for Indian corn," while that at the straits was poor, where the proposed Fort Michilimackinac was to be built (Innis 1962:61). Then and later Detroit was a major source of corn. John Askin's inventory of his estate notes that he had "138 Bushels of Detroit Corn @ 20 Shillings" and "about 11 Bushels of Arbra Croch Corn @ 12 Shillings" (Armour and Widder 1978:218). In the John Askin papers there are frequent references in the years 1778 to 1786 to the importation of corn from Detroit to the old fort on the south side of the straits and to the later one on Mackinac Island (Quaife 1928:74, 75, 80, 104, 263). Some of this was partly processed for voyageur use by being "hulled."

In a letter dated April 28, 1778, which John Askin wrote to a correspondent in Montreal, he stated, "I shall send a Vessell to Millwakee in search of Corn. I have 150 Bushells already there & hope for more" (Quaife 1928:75).

Corn at times came from as far away as Niagara. In a petition dated October 4, 1784, the Northwest Company, seeking release from regulations barring private vessels, requested: "Such provisions especially Indian corn must necessarily be purchased about Niagara, at Fort Erie, when it cannot be had in the settlement of Detroit, and all the company wishes for is to have the preference in the transport to Michilimackinac in the King's vessels, if ready in point of time; if not, to suffer a private vessel to be commanded by a King's officer if his Excellency thinks fit, to take their provisions forward from Fort Erie or Detroit, and thence to Michilimackinac" (Innis 1962:183).

7 Corn from Fort Michilimackinac, A.D. 1770–1780

Leonard W. Blake and Hugh C. Cutler
Missouri Botanical Garden

(Written in 1968)

Leonard Blake's Comments, 1999
   The corn described in this report was received from Dr. James Brown, then of Michigan State University, who was conducting excavations at the site. The samples were recovered from the university's 1967 excavations at the rear of the houses used historically by French inhabitants.
   Corn that was shipped in to Fort Michilimackinac almost invariably consisted of shelled corn, that is, kernels stripped off the cob, for ease in transport. The recovery of charred corn kernels in and around the houses was very common. These 1967 excavations were one of the very few instances in which cobs were recovered and thus are of unusual interest.

   Fort Michilimackinac was and still is near the northern limit of corn agriculture. Cartier found Indians growing corn at the present site of Montreal in 1535 (Tooker 1964:3). This is at about the same latitude as the Straits of Mackinac. Corn was grown still farther to the north along the Missouri River: Vérendrye found the Mandan villages growing large crops of corn during the eighteenth century in what is now North Dakota (Will and Hyde 1968:60). Present-day farmers grow corn mostly for silage in that area and few attempt to produce mature ears. The principal limiting factors in growing corn in high latitudes are summer temperatures and the length of the frost-free growing season. The last was of less importance to the Indians because of their practice of soaking the kernels before planting to hasten sprouting and because they harvested as soon as the kernels were firm but not necessarily dry (Kinietz 1965:16). The growing of corn near the straits and elsewhere in the somewhat unfavorable environment of the Upper Great Lakes was probably stimulated by the scattering of the Hurons and Neutrals by the attacks of the Iroquois around the middle of the seventeenth century. The Hurons were heavily dependent on corn. They were capable corn farmers and had varieties well adapted to a northern climate. These things are repeatedly mentioned in the early historic accounts dating back to Champlain's visit of 1615 (Tooker 1964:60) and are confirmed by archaeological excavations for even earlier dates.

The corn cob samples recovered in Michigan State University's 1967 archaeological excavations are from the rear of the houses of French inhabitants of A.D. 1770 to 1780. They are essentially a pure form of a race of corn variously called Northern or Northeastern Flint. This was the dominant kind of corn grown from southern Canada, New England, and New York across the lake states to the plains in late prehistoric and historic times until after 1850. In this period it appeared mixed with other kinds of corn on the western part of its range on the plains and to the south. This hardy and early maturing race is one of the ancestors of our present corn belt corns. The ears are described by Brown and Anderson (1947:2) as "characteristically long and slender with 8 to 10 rows of wide, crescent-shaped kernels. The cob is strong and proportionately large, particularly toward the base, and the shank or ear-stalk is thick and well developed. Frequently the base is noticeably larger than the rest of the ear."

Although there are a few large cobs in the sample, the median size is below that of collections of similar corn from other locations. It is known that the soil at Fort Michilimackinac is poor and sandy, but this is not believed to be the principal cause of the substandard cobs. Peter Pond (1900:328), a Connecticut Yankee, who stopped several times at the fort in the period, wrote, "The Land about Macinac is Vary Baran—a Mear Sand Bank—but the Gareson By Manure Have Good Potaters and Sum Vegetables." Peasants in France and their eighteenth-century descendants in Canada and the Mississippi Valley had a reputation for skill and care in cultivation of their kitchen gardens. Such people would be almost certain to offset the effect of poor soil to the extent that they were able by regularly fertilizing with compost, kitchen waste, and animal manure, when available. Without fertilization, it is doubtful that corn could have been grown at all.

The climate of the straits region is not a favorable one for corn agriculture, yet carbonized corn kernels were found in archaeological excavations by the University of Michigan on nearby Bois Blanc Island, which lies eight miles to the east. This corn was dated by the carbon 14 method within the period of A.D. 1070 to 1320. It was not possible, however, to ascertain whether it was grown on the spot or brought in from another place (Yarnell 1964:17, 42, 123).

According to the United States Department of Agriculture Yearbook of 1941, the average frost-free season at Mackinaw City, over an observed period of forty years, was 148 days. This is not overly long, but it is long enough to mature a crop of Northern Flint corn. The charred corn cobs from this site appear to be from fully grown ears. We have cobs that are larger on the whole from other sites where the average growing season is even shorter.

Average June rainfall, when the corn plants would be making their growth, is less than at most of the other sites listed in Table 7.1, but additional moisture from lake fogs, which do not show up in Weather Bureau precipitation figures, would be available to the plants. Fog would also reduce water loss. Leonard and Martin (1963:167) state that "the period of highest water requirement [of corn] appears to be from tasselling to kernel formation." Here this would be in August and rainfall is normally greater in that month than at the locations of two sites along the Missouri River in a similar latitude that produced larger cobs (Table 7.1).

Northern Flint tends to "sucker," that is, to send out shoots on which are sometimes borne smaller "nubbin" ears. The low average cob size does not appear to be the result of a large proportion of ears from suckers in the sample. None is deformed and there is no sharp break in the sizes of the cobs from the smallest to those of median size, as would be the case if sucker ears were included in the collection.

Because soil, length of the growing season, rainfall, or sucker ears do not appear to account for subaverage cob size, temperature and sunlight during the growing season should be considered. Mean temperature for the month of July at Mackinaw City is 66.8 degrees F. This is the lowest of any of the site locations listed in Table 7.1. Most plants show little growth at temperatures below 65 degrees F. Growth of corn is highest at air temperatures of 77 to 86 degrees F, according to Leonard and Martin (1963:166). Low average temperatures and the occasional shielding of the growing plants from the sun by lake fogs would slow down photosynthesis and hence stunt growth.

When the disadvantages of low summer temperatures and reduced sunlight are considered, it appears surprising that nearly 20 percent of the thirty-five corn cobs in the sample are near or above normal size. The size of these cobs may be a result of unusually favorable circumstances or the skill of the French gardeners, or the corn may have been brought in from some other place and not grown in the immediate vicinity of the fort at all. We would like to consider this possibility.

From the journals of the time, it is known that corn was often in short supply at Michilimackinac. John Askin, a trader there, speaks in his diary (Quaife 1928:75) of sending boats to Detroit and even on one occasion to Milwaukee to purchase corn. According to his account, this was usually bought shelled from the cob or "hull'd," as he expressed it. The purpose of buying shelled corn was, obviously, to bring the most that the limited carrying capacity of his boats would hold. It seems entirely possible that some unshelled corn would be occasionally brought back when sufficient shelled corn was not available at the port of call. Detroit, from which most corn was shipped, has a mean July temperature of 73.1 degrees F and Milwaukee 70.1 degrees F. Either place could be the source of the few large cobs.

Corn also came to Michilimackinac from the Ottawa settlement of L'Arbre Croche, some thirty miles to the southwest. Peter Pond (1900:328) said of this

Table 7.1. Comparative Data on Northern Flint Corn from Selected Archaeological Sites

Site	Date, Culture, Tribe	July Mean Temp. (°F)	Frost-Free Growing Season (days)	Average Monthly Rainfall (in.)			No. Cobs	Mean Row No.	Mean Cupule Width (mm)
				June	July	August			
Ft. Michilimackinac Cheboygan County, Michigan	French inhabitants, Late British occup. A.D. 1770–1789	66.8	148	2.65	2.69	2.76	35	8.1	8.5
BdGul-122, Sopher site Near Orilla, Ontario, Canada	Protohistoric Huron-Iroquois ca. A.D. 1570–1590	68 (est.)	Not available	2.64	2.8 (est.)	2.8	9	8.4	9.0
BfFu 1, McIvan site 40 miles S. Ottawa, Ontario, Canada	Prehistoric Onondaga ca. A.D. 1500	70.2	152	3.24	3.19	2.87	37	9.0	9.4
32Bl8, Double Ditch site Burleight County, North Dakota	Heart R. Focus Protohistoric Mandan A.D. 1700–1725	70.9	140	3.22	2.22	1.64	29	8.6	9.3
Alhart site Monroe County, New York	Prehistoric Iroquois ca. A.D. 1400–1600	71.1	162	2.94	3.13	2.75	14	8.1	12.8
39Ca6, Bamble site Campbell County, South Dakota	Arikara ca. A.D. 1690–1750	72.4	123	3.1	2.04	1.63	37	8.3	11.8

Site	Description							
Sheeprock Shelter Huntington, Pennsylvania	Shank's Ferry and Susquehanock Most A.D. 1550 & later	72.7	147	4.04	4.09	4.07	1,714	8.3	11.3
11Ri81, Crawford Farm Rock Island County, Illinois	Historic Sauk ca. A.D. 1790–1810	76.6	182	4.25	3.23	3.59	1,300	8.3	9.7
23Ve4, Coal Pit site Vernon County, Missouri	Historic Little Osage ca. A.D. 1775–1820	76.6	186	5.2	3.44	3.55	328	8.9	10.4

All samples except that from Sheeprock were carbonized and median cupule width has been increased 20% to adjust for shrinkage that occurs when corn cobs are carbonized. Cob size is a function of width of units of the cob (cupule width) and the number of rows of these units.

village: "Thare is Sum lndan Villeges twenty or thirty Miles from this Plase whare the Natives Improve Verey Good Ground. Thay Have Corn Beens and meney articles which thay youse in Part themselves and bring the Remander to Market. The Nearest tribe is the Atawase [Ottawa]."

Around the corner of the lake from the fort, on the west side of the Michigan peninsula, would be a better place to grow corn than at the straits. The soil was good, as Pond indicates. The prevailing south and southwest winds that blow during the summer months pile up the warmed surface water of Lake Michigan toward the shore and raise the temperature in the later part of the summer. Rainfall during the growing season is almost the same as at the fort, but fogs are less prevalent and the hours of summer sunshine are greater.

Not only was the corn of the northern Indians predominantly of the same race from New York to the Great Lakes, but also the methods of harvesting were much the same. Most of the corn was gathered after the ears were fully formed, though sometimes before the kernels were fully dry (Kinietz 1965:370). A quotation from Sagard's "Le Grand Voyage" describes the harvest as practiced by the Hurons near the lower end of Georgian Bay in 1623–1624 (Kinietz 1965:17):

> After that they gather it and tie it by the turned-down leaves; they do it up in packets, which they hang all along the length of their cabins. . . . The grain being well dried and fit to pound, the women and girls shell, clean, and put it in their large vats or casks for this purpose.

It seems a reasonable possibility that some of the corn brought to the fort by the Ottawa to trade might have been still on the cob. This seems particularly likely in view of the fact that newly harvested corn would be the first item of consequence that they had to trade after what must have been to them a long summer.

The carbonized corn found behind the houses of the French inhabitants probably was grown by them nearby, but all or part could have been brought in from other places near or far away.

Explanation of Measurements and Terms Used

Gross cob measurements, made on the entire cob after grains have been removed, are very inaccurate. The glumes are frequently broken or may be so soft that a definite measurement is difficult to make. Frequently only fragmentary cobs, with the glumes broken or entirely gone, are recovered from sites. It is best to make measurements of a single unit of the main axis of the cob, the rachis. Such a measurement, cupule width, devised by Nickerson (1953), has often been used. A cupule is the pocket in the central axis of the cob in which a pair of spikelets, each usually producing a single grain, is borne. Width is measured across the cupule, from one margin to the other at right angles to the longitudinal axis.

Cob size is the function of the width of units of the cob (cupule width) and the number of rows of grains. Because a pair of grains is borne in each cupule, the number of cupules visible in a cross-section is half the rows of grains (Cutler 1966:11–12).

Cob rachis diameter is the diameter of the hard part of the cob, exclusive of the glumes. It may be calculated mathematically, with reasonable accuracy, when the number of rows of grains and the cupule width are known, as the cross section of an eight-rowed cob is a rectangle, that of a ten-rowed cob is a pentagon, etc.

Measurements on all carbonized cobs have been increased 15 percent to adjust for shrinkage of 10 to 20 percent that occurs when corn cobs are carbonized.

8 Corn from the Waterman Site (11R122), Illinois

Leonard W. Blake
Washington University

(Written in 1972 and 1997)

Leonard Blake's Comments, 1999

Samples of corn from the Waterman site were originally received during the 1970s from Dr. Margaret K. Brown, superintendent of the Cahokia site from the 1970s through the 1980s, and a report was written. At a later date, more samples were received. I rewrote the report using this new information, and added information on a sample from nearby Fort de Chartres, from Dr. M. D. Thurman of Ste. Genevieve. He estimated that it had been grown at a slightly later date than that from Waterman. Estimates of dates and identification of cultures of all sites listed are those of the excavators.

Historically, the Waterman site was the village inhabited by the Michigamea tribe, a member of the Illinois Confederacy. It is located about two miles north of Fort de Chartres. The village was destroyed in 1752 in a surprise attack by the Fox Indian tribe and their allies. Dr. Brown excavated burials of victims of this attack, as well as numerous smudge pits.

All of the material from the Waterman site (11R122) came from "small, circular straight-sided pits corresponding to Binford's (1967) description of hide smoking pits" (Letter from Dr. Margaret K. Brown, May 21, 1972). Occupation of the site was estimated at about twenty years, between A.D. 1750 and 1770. Information was furnished that makes possible the probable allocation of many of the corn cobs recovered to individual households or groups. For example, it was stated that "Pits 41A, B, C are a unit located close to each other and probably belong to a single house.... Pits 42, 43, 44 also a unit" (Letter from Brown, May 21, 1972). In addition, several individual pits contained enough cobs for comparative purposes. Table 8.1 shows that the corn from the different sources is similar, but not identical. Corn from Features 42 and 44 is notable for having somewhat larger cobs than those from the other units.

The corn from Waterman is different from that found on sites of most historic northern Indians, including other members of the Illinois Confederacy.

Table 8.1. Corn Cobs from Different Locations on the Waterman Site

Corn Cobs	Fea. or House No.							All Other Corn Cobs
	F. 80	F. 57	House 2	F. 62	F. 42, 44	F. 18	F. 41B, C	
Percent 8-rowed	21	26	29	49	52	51	52	32
Percent 10-rowed	67	52	56	27	42	39	42	46
Percent 12-rowed	9	2	13	18	6	10	6	22
Percent 14-rowed	3	—	2	6	—	—	—	2
Mean row no.	9.9	9.8	9.7	9.6	9.4	9.2	9.1	9.8
Median cupule width (mm)*	7	7.2	7.4	7.8	8.5	**	7.6	7.0
No. cobs	34	19	95	33	94	39	48	59

*Cupule width is the distance across the entire pocket in which a pair of grains and their spikelets is borne. It is a measure of the size of the cob. For example, a 10-rowed cob will have a larger diameter than an 8-rowed cob with the same cupule width, and a 12-rowed cob will have a still larger diameter.
**Not measured.

This is not surprising in view of the Michigamea's frequent contacts with other groups to the south in the historic period (Temple 1958:18, 29, 32, 33, 34, 36, 39, 46).

The kind of corn grown by Indians in the northeastern United States at this time has been called Northern Flint (Brown and Anderson 1947), but more recently it has been termed Eastern Eight Row (Cutler and Blake 1976:5), for the kernels may be flint, flour, or sweet. The ears usually have eight rows of kernels, although they may have ten, or, rarely, twelve. The corn is cold resistant and early maturing. It was dominant throughout the northeastern United States from about A.D. 1200 into historic times.

The only sample of corn from a site occupied by Indians of the Illinois Confederacy that we had seen before is from the Zimmerman site, which probably was the 1673–1680 Kaskaskia village visited by Marquette (Brown 1975:1–2). It consisted of twenty Eastern Eight Row corn cobs, fourteen of which were eight-rowed and six of which were ten-rowed (Blake and Cutler 1975). To present a larger and more typical collection for comparative purposes, I include information on corn from the Crawford Farm site (11Ri81) of the Sauk Indians ca. A.D. 1790–1810 in Table 8.2. This sample is almost identical in composition to a large sample from the Rhoads Kickapoo site (11Lo8) of about the same time period. Corn from these sites is considered typical Eastern Eight Row.

Table 8.2. Comparisons of Corn Cobs from the Waterman Site (11R122) with Those from Other Indian Sites of about the Same Time Period

Site name	Waterman	Fort de Chartres	Upper Nodena	Crawford Farm	Nuyaka
Site number or location	11Rl22	Randolph Co., Illinois	3Ms4	11Ri81	Horseshoe Bend, Ala.
Excavator	M. K. Brown	M. D. Thurman	D. F. Morse	E. B. Herold	C. Fairbanks
Culture or historic tribe	Michigamea	Indian or Euro-American	Mississippian	Sauk	Upper Creek
Approx. date A.D.	1750–1770	1772–1800	1400–1600	1790–1810	1777–1813
Total corn cobs	421	168	98	1308	28
Percent 8-rowed	40	24	11	86	4
Percent 10-rowed	47	46	41	12	21
Percent 12-rowed	12	25	38	2	39
Percent 14-rowed	1	4	9	—	25
Percent 16-rowed	—	1	—	—	11
Mean row no.	9.5	10.2	10.9	8.3	12.4
Median cupule width (mm)					
8-rowed cobs	8.1	8.0	7.6	8.5	8.8
10-rowed cobs	7.3	7.8	7.4	8.4	9.0
12-rowed cobs	6.9	7.0	6.2	7.5	8.3

14-rowed cobs	6.2	6.0	5.2	—	7.3
16-rowed cobs	—	5.0	—	—	7.0
No. cobs measured	133	168	98	243	28
Median cupule width for all cobs measured	7.5	7.4	6.9	8.5	8.2

All cobs are carbonized and values not adjusted for shrinkage.

The corn from the Waterman site has been heavily influenced by this kind of corn, as 40 percent is eight-rowed. It will be seen that it has also been influenced by another kind of corn, for 60 percent of the cobs are from ears with more than eight rows. This may be a result of the Michigamea's frequent visits south of the Ohio River (Temple 1958).

Along the Mississippi River, between the mouth of the Ohio and the Arkansas River and extending west through Oklahoma, the corn was not Eastern Eight Row. The kinds grown there had small cobs with higher row numbers. These kinds of corn were present prehistorically and continued in the area into the historic period (Blake 1986). A good example may be seen in corn recovered from the Upper Nodena site (3Ms4) in northeastern Arkansas, data from which are shown in Table 8.2.

There seems to be some possibility that the corn at Waterman may have been influenced by still another kind of corn, although the probability appears less strong than in the case of the corn from the nearby Fort de Chartres. We are grateful to Dr. Melburn D. Thurman for the opportunity to see the corn that he recovered from a single large pit at the fort. He has said (personal communication) that there is evidence that this corn was deposited by Indians some time after the fort was abandoned by the whites, approximately between A.D. 1772 and 1800. Comparative figures on this corn are found in Table 8.2.

The kind of corn mentioned above is dent corn, varieties of which became the traditional corn of southern white farmers in the late eighteenth and early nineteenth centuries. Dent corn, which originated in Mexico (Brown and Anderson 1948:264), is distinguished by the dent or shallow groove in the top of the dried kernels caused by the arrangement of soft and hard starch in the interior. There are a number of varieties, most, though not all, with many rows of kernels. The kinds that resulted from a mixing of Mexican varieties with the corn already present in the southeastern United States have been summarized by Brown and Anderson (1948), who called them "Southern Dent Corns." To the best of our knowledge, these corns were not grown prehistorically in the eastern United States. Although the Spanish began importing corn from Mexico from the beginning of settlement at St. Augustine in 1565 (Lyon 1976:179), and a single Mexican corn cob was recovered from a sixteenth-century site in South Carolina (Cutler 1980), Mexican dent varieties were not grown in the eastern United States until somewhat later. There is a historic reference to dent corn in Virginia in 1679 (Ewan and Ewan 1970:140) but we have seen no positive archaeological evidence of its use in the Southeast before the eighteenth century in corn from the Upper Creek village of Nuyaka in Alabama (Table 8.2).

The corn from Fort de Chartres shows a substantial decline in the proportion of eight-rowed cobs and an increase in the proportions of twelve- and fourteen-rowed cobs when compared with that from Waterman. These changes seem too large to have occurred without something new having been added, particularly in view of the short time difference in the occupations of the two sites. It seems reasonable to suspect that the changes in the corn from the fort may be a result of some mixing with the Southern Dents, which were becoming widespread at about this time.

9 Plant Remains from the Rhoads Site (11L08), Illinois

Leonard W. Blake and Hugh C. Cutler
Missouri Botanical Garden

(Written in 1974)

Leonard Blake's Comments, 1999

The Rhoads site, in Logan County, Illinois, was excavated by an archaeologist employed by the Illinois State Museum, as an archaeological salvage project, made necessary by the needs of the Illinois Department of Transportation. Our report was to be an appendix to a final report by the museum's archaeologist. We received some information about the history of the site, some data from the field notes, and all plant remains recovered. I wrote this report and turned it in to the Illinois State Museum, but no publication followed. I believe that information in this report is still valid and not outdated, although form of presentation and emphasis might not be identical today.

Plants from archaeological sites can provide valuable information on the lives of past inhabitants. The kinds of wild plants indicate what portions of a region were exploited most thoroughly and sometimes suggest the time of year when the site was occupied. Cultivated plants are especially useful. Sometimes they can indicate the degree of sophistication of plant selection and agricultural techniques, and by comparisons of the kinds of plants grown with those of their neighbors, provide a measure of the amount of cultural interchange.

The Rhoads site is important because it lies in a region that was undergoing great change during the site's period of occupation. Corn was becoming more important for food, and as a medium of exchange. New kinds of corn and other plants were introduced by waves of people moving into or through the region. The carefully controlled excavation of the Rhoads site recovered enough specimens, most of them preserved by carbonization, so that a series of studies and comparisons could be made in addition to simple identification.

Some form of flotation, as defined by Struever (1968), or the more rapid but sometimes more destructive method called "water screening," is becoming standard practice in American archaeology. Water screening usually consists of placing excavated dirt on a screen, or screens, and squirting it with a hose. From light soils recovery may be good and damage minimal, but with heavy clays, or gumbo, fragmentation may be great. Flotation usually produces more undamaged, or lightly damaged, carbonized plant material. Either one is a great step forward over the practices of just a few years ago when recovery of this class of remains was sporadic and unorganized.

The extent of recovery of fragile, carbonized, botanical material may vary widely because of chance differences in abundance, and because of differences in the proportions of excavated dirt processed. The methods used and the degree of care and understanding of the individual excavator and processor are also factors. Preservation of any archaeological remains is by chance, to some extent, and the odds for charring, usually necessary for the preservation of plant remains in a humid climate, are even less. Some remains are more likely to become charred than others. For example, ethnological accounts indicate that corn was often parched (Kinietz 1965:175; Waugh 1916:88). This would appear to increase chances that corn would become overparched, that is, carbonized, more than other kinds of plant foods that were processed by other methods, such as boiling. The use of corn cobs for fuel or smudges could be expected to increase the frequency, and consequently the recovery, of carbonized cobs. The hard parts of carbonized corn cobs are nearly indestructible, as are some nut shells, plum stones, etc., but seeds of beans and squash are extremely fragile when carbonized, and so may be underrepresented, even when the most careful procedures are followed.

Hackberry trees grow on disturbed soil. People have sucked the sweet pulp from the outsides of the seeds since the days of Peking Man (Coon 1962:431), but who can say which seed found on an archaeological site was savored by an Indian, and which fell off a tree growing on a midden deposit. Animals and birds too like to eat some of the same things that people do. The presence in quantity or in many parts of a habitation site of such things as plum pits or pawpaw seeds probably means that people had been eating the fruit, but an occasional occurrence might be incidental burning. These matters indicate a need for caution in viewing statistical results from analysis of carbonized plant remains from archaeological sites.

On the positive side, plant remains can provide several kinds of information. As more collections are available, and techniques are improved, their usefulness increases. The frequency of certain kinds of plants may indicate their relative importance to the inhabitants, with some exceptions such as are noted above. Both wild and cultivated plants can give clues as to the time the site was occupied, and, sometimes, food preferences or prejudices. Wild plants may indicate what part of the environment was being exploited. Corn and other cultivated plants dependent on human activity usually indicate that agriculture was practiced by the people who lived on the site. Similarity of the corn to that on other sites may indicate trade or contact. Cultivated plants have been changed by human manipulations, but they cannot be independently invented, for the changes are limited by the plant's genetic inheritance. The presence of seeds of an Old World plant, which were difficult to transport under primitive conditions, could indicate not only contacts with Europeans, or their agents, but also an active, as well as passive, acceptance of things foreign. The Indians would have had to grow the plant themselves for the seeds to be present in more than one or two locations.

Corn

Carbonized corn cobs or corn grains were found in thirty-three, or 62 percent, of the fifty-three Class 1 pits excavated, and corn cobs were present in fifty-eight, or 67 percent, of all Class 2 "smudge" pits. The collection is considered a representative one, for a close watch was kept for plant remains during excavation and the entire contents of thirty-nine of the forty-one larger, Class 1, Subclasses A and B, pits and two of the Subclass C pits were water screened. In addition, a sample was water-screened from ten of the remaining twelve Class 1 pits.

The procedure in handling corn cobs was to count the number of rows of grains and to measure cupule width. A cupule is "the pocket in the central axis of the cob in which a pair of spikelets, each usually producing a single grain, is borne. Width is measured across the cupule from one margin to the other, at right angles to the longitudinal axis" (Cutler 1966:12). Experience has shown that these simple measurements may be made more quickly and accurately, and more easily, on cob fragments than gross cob measurements, and that they provide fairly sensitive indicators of differences in corn through time and space. Our experiments have shown that corn cobs shrink approximately 15 to 25 percent when carbonized. No adjustment has been made for shrinkage in the attached measurements of corn cobs. Some estimated adjustment should be made if this carbonized collection is to be compared with an uncarbonized one, but when compared with another large, carbonized one, such adjustments are not considered necessary.

Not every corn cob fragment recovered from the site was measured. An effort was made to get a sampling from most of the features. Where the number of cobs was large, they were laid out in rows, and every second, third, or fourth cob, depending on the size of the sample, was picked out for measurement in an effort to avoid bias in selection. If only fragments or loose cupules were recovered from a feature, then an effort was made to get measurements from fragments that did not appear to be from parts of the same cob.

The corn from this site belongs to the race that has been called Northern or Eastern Flint (Brown and Anderson 1947) or, more recently, Eastern Eight Row

(Cutler and Blake 1976). This race of corn has a large cob with an expanded butt, usually bearing eight, but sometimes ten, or, more rarely, twelve rows of grains, which are crescent-shaped; that is, they are wider than long. These grains are usually hard flint or soft flour, rarely sweet. The plant is cold and drought resistant and requires a relatively short growing season to reach maturity. This hardy race of corn was dominant throughout the northern and eastern United States from about A.D. 1200 to 1850. The most extreme forms have come from about A.D. 1500 from the northern tier of states from New England to Minnesota, grading into varieties of the same race with higher average row numbers in the south and west.

Eastern Eight Row corn from the northern states is quite uniform, and was for some time. An indication of this may be seen by comparing measurements on 123 corn cobs from the Whittlesey focus South Park site of shortly after A.D. 1500, near the modern city of Cleveland, Ohio, with those from the Rhoads site of several hundred years later. The mean row number of carbonized cobs from South Park is 8.3, and the median cupule width is 8.5 millimeters (Cutler and Blake 1976:67). The mean row number of 262 cobs measured from the Rhoads site is 8.4, and the median cupule width is 8.3 millimeters.

In counting row numbers, it is our practice to list abnormal four- and six-rowed ears as eight-rowed in statistical summaries, as they usually are from plants that under more favorable conditions would bear eight-rowed ears. Such cobs, when numerous, may give indication of unfavorable growing conditions. The number of four- and six-rowed cobs counted was proportionally small, amounting to less than 4 percent of the total. Table 9.1 contains comparisons with three other historic Indian sites in the Midwest that overlap the Rhoads site in time of occupation. Corn from the Crawford Farm site in Illinois (11Ri81), which was occupied by the Sauk and Fox, is nearly identical to that from the Rhoads site, both in proportions of row numbers and in proportions of cob sizes, as indicated by cupule width. This is not surprising in view of the close relations and similarity of language between these tribes. According to Temple (1958:166–67), some Kickapoo joined the Sauk on Rock River in 1812, where the British were distributing gunpowder. They "returned to the village on Kickapoo Creek near Lincoln after the war of 1812."

Corn from the other two sites used for comparison here is also considered to be Eastern Eight Row, but each shows some influence of mixture with other kinds of corn. Corn from the Little Osage Coal Pit site in Missouri (23Ve4) has a higher mean row number (8.9) and larger cobs than that from the Rhoads site. This may be due in part to a longer growing season. The site is located in a lower latitude in the southern half of western Missouri. It is also possibly a result of some hybrid vigor from a slight mixture with the productive Southern Dent corns, which began to come out of Mexico into the Southwest and parts of the South about A.D. 1700.

Table 9.1. Comparisons of Corn Cobs from the Rhoads Site (11Lo8) with Selected Midwestern Sites

Site name	Rhoads	Crawford Farm	Coal Pit	Waterman
Site number	11Lo8	11Ri81	23Ve4	11R122
Approximate period of occupation	A.D. 1760–1820	A.D. 1790–1810	A.D. 1790–1820	A.D. 1750–1770
Historic tribe	Kickapoo	Sauk & Fox	Little Osage	Michigamea
Total no. of cobs counted	262	1308	328	421
Percent 8-rowed	85	86	63	40
Percent 10-rowed	12	12	30	46
Percent 12-rowed	3	2	6	13
Percent 14-rowed	—	—	1	1
Mean Row No.	8.4	8.4	8.9	9.5
No. cobs cupule width measured	262	226	292	133
Cupule width = 8.6+ mm (%)	43	43	59	21
Cupule width = 7.6–8.5 mm (%)	28	34	20	26
Cupule width = 6.6–7.5 mm (%)	21	19	14	30
Cupule width under 6.6 mm (%)	8	4	7	23
Median cupule width (mm)	8.3	8.4	9.0	7.5
Cobs from smudge pits*	174	495	246	All cobs are from smudge pits
Mean row no.	8.4	8.5	9.0	
Median cupule width (mm)	8.3	8.4	9.2	

*Included in the figures above.

Corn from the Waterman site in Illinois (11Rl22), which was occupied by the Michigamea, a member of the Illinois Confederation, has a much higher average row number (9.5) than corn from the other sites compared here, and the median cupule width is smaller. This corn is somewhat like that grown by the Indians in northeastern Arkansas in late prehistoric times. It is about what one would expect from a mixture of southern forms with varieties grown farther north. As early as A.D. 1687, Henri Joutel found the Michigamea several days' journey below the mouth of the Ohio River. Between then and the end of occupation of the site, there was movement back and forth between Arkansas and the Illinois country (Temple 1958:29, 32, 34, 44).

Dr. Walter Klippel, the archaeologist in charge of excavations at the Rhoads site, in unpublished notes (Sec. 4, Archaeology—Features), described and defined Class 2 features as "smudge" pits. Mention has also been made of the arguments of Binford (1967, 1972) and Munson (1969) based on ethnographic use of such pits for smoking hides or for waterproofing pottery. At all four of the sites listed in Table 9.1, such pits were abundant. At the late date of all of these sites, it is to be expected that the making of native pottery would be on the decline, if it had not already ceased, following the introduction of metal kettles through trade. The principal item of trade used to obtain kettles and other desirable goods in the Midwest was deerskins. Amos Stoddard, who took possession of Upper Louisiana for the United States in 1804, estimated that in the previous fifteen years an average of 1,158,000 pounds of deerskins had been traded annually, as against a total of 80,650 pounds for all other pelts (Stoddard 1812:297). The value of deerskins was about 31 percent of the total average fur trade. Although the evidence is circumstantial, a decline in pottery making, a large number of smudge pits, and the economic importance of deer hides suggest deer hides may have been smoked over Class 2 pits. Whatever their true purpose, the smudge pits have produced a large sample of corn cobs. At Rhoads, the cobs from smudge pits are similar to those from other parts of the site. Cobs from smudge pits at Crawford Farm and Coal Pit are slightly larger than those from other parts of the site.

Carbonized corn grains are less reliable than carbonized corn cobs for measurements, for they tend to swell under the action of heat and to become distorted. When corn grains are charred on the cob, the pressure of the surrounding grains lessens distortion, and useful measurements may sometimes be made. Corn grains occurred less commonly than corn cobs at this site. They were nearly all quite large, ranging in width from just under 7 millimeters to nearly 13 millimeters. Most were over 10 millimeters wide. All that were whole enough to observe were wider than long and crescent shaped, as is typical of corn of the Eastern Eight Row race.

Beans

Carbonized domesticated common beans (*Phaseolus vulgaris*) were recovered from thirteen of the Class 1 pits. This is equivalent to just less than 25 percent of the fifty-three pits of this class that were excavated. In our experience, recovery from such a large proportion of excavated units is unusual. It could indicate that beans were abundant, and it certainly indicates care on the part of the archaeologists excavating and processing dirt from features because, as pointed out above, carbonized beans are fragile. Carbonized beans are also difficult to classify. Color, which is often diagnostic (Kaplan 1956:201, 202), has been destroyed. Carbonization also usually destroys the skin, so that only halves of the seed remain and dimensions are distorted. We have found through experimental charring that a wide variation in shrinkage or expansion may occur. Within broad limits, however, most carbonized beans we have seen from sites in the eastern United States fall within ranges indicating some uniformity in size and shape. This is in contrast to the southwestern United States where diversity is wide. With the exception of three large beans from Feature 17, other beans from this site are similar in size to those from the Zimmerman site (11Ls13) on which the historic Kaskaskia lived between A.D. 1673 and 1691, and similar to those from an earlier Mississippian site near Kansas City of about A.D. 1000–1200 (see Table 9.2).

Size ranges of carbonized beans from the Rhoads site are shown in Table 9.3. Nearly all of the beans except the three large ones, and all but a few that we have seen from the eastern United States, appear to be small when compared with beans from the southwestern United States, even when generous allowances are made for shrinkage. It has occurred to us that possibly some of the beans that are being recovered from humid Midwestern sites are small ones from the ends of the pod. Beans are too valuable a source of protein to waste, but if beans were shelled, and the pods then thrown into a fire, it would be possible for the pods to burn, and a few unnoticed beans in the ends to be carbonized. This is frankly speculation, however.

Most of the beans from this site and those we have seen from historic and prehistoric sites in the Midwest are similar to some of the many kinds of small bush and vining beans described in *Beans of New York* (Hedrick et al. 1931). Most of these beans were obtained from the Indians of the region, and many other kinds are still grown by the Iroquois in New York (Waugh 1916:103–8, 215, plate 34).

Dimensions of the large beans in Feature 17 are 16.7 by 8.5 by 6.0 millimeters thick, 16.4 by 8.1 by 6.3 millimeters, and 15.4 by 8.7 by 4.5 millimeters. These are the largest beans we have seen from archaeological sites in the eastern United States, and are large even when compared with uncarbonized beans from the Southwest. They resemble a large bean variety that is still grown in the Midwest under the name "Red Kidney Bean" (Hedrick et al. 1931:plate facing p. 80).

Table 9.2. Beans from the Rhoads Site (11Lo8) Compared with Those from Other Sites

Site	Excavator	Culture	Est. Date (A.D.)	Specimens Measured	Range (mm)		Median (mm)		Ratio (mm)
					Length	Width	Length	Width	L/W
11Lo8	W. E. Klippel	Historic Kickapoo	1760–1820	41*	16.7–7.0	8.7–3.8	9.4	5.7	1.65
11Lo8	W. E. Klippel	Historic Kickapoo	1760–1820	38**	12.7–7.0	7.2–3.8	9.4	5.6	1.68
11Ls13	M. K. Brown	Historic Kaskaskia	1673–1671	38	12.3–7.2	7.6–3.3	9.5	5.0	1.90
23Pl4	P. J. O'Brien	Mississippian	1000–1200	33	12.6–8.0	7.1–4.7	9.6	6.0	1.60

All specimens carbonized; most are halves of beans. Values not adjusted for possible shrinkage.
*All specimens that were measured.
**Less three large beans from Fea. 17.

Table 9.3. Beans (*Phaseolus vulgaris*) from the Rhoads Site (Carbonized)

Half Beans (with skin burnt off) (length & width in mm)		Whole Beans (length & width in mm)
12.7 × 7.2	9.3 × 5.6	16.7 × 8.5
12.1 × 6.3	9.0 × 5.9	16.4 × 8.1
11.6 × 6.4	9.0 × 4.9	15.4 × 8.7
11.0 × 7.0	8.7 × 5.7	11.3 × 6.1
10.7 × 6.6	8.6 × 5.3	10.4 × 5.3
10.2 × 6.4	8.3 × 4.5	10.1 × 6.2
10.1 × 6.8	8.2 × 5.8	9.9 × 5.5
10.0 × 6.2	8.2 × 5.4	9.4 × 6.1
10.0 × 6.1	8.1 × 5.0	9.2 × 4.9
9.9 × 6.2	8.0 × 5.4	8.7 × 5.4
9.7 × 6.4	7.9 × 5.0	7.9 × 4.0
9.7 × 5.4	7.9 × 4.4	7.3 × 4.5
9.5 × 6.1	7.5 × 4.1	
9.4 × 5.3	7.4 × 5.0	
	7.0 × 3.8	

Squash

Squash seed or cotyledons from the inside of the seed were recovered from fourteen, or slightly more than 26 percent, of the Class 1 pits. Measurable specimens were present in eleven of these. Some uncarbonized seeds were present in Features 2, 17, 41, 46–70, 118, and 163. Some of those from Feature 2 and all of those from Feature 17 were stained green, and possibly owe their preservation to the action of copper salts from objects of copper or brass. Many of the other uncarbonized seeds had been deeply buried. There is no reason to believe that they are intrusive. Seeds of cucurbits such as squash, gourd, and watermelon show remarkable resistance to decay when buried under favorable conditions. A large number of uncarbonized squash seeds (*Cucurbita pepo*) of several different kinds were recovered from a Mississippian pit, dated around A.D. 1000. The pit, which was excavated by Charles Bareis of the University of Illinois-Urbana, was under Mound 51 near Monk's Mound in the large Cahokia village site near St. Louis (Cutler and Blake 1976:23).[1]

Squash seeds from the Rhoads site, which were measured, ranged from 7.9 to 18.5 millimeters in length (see Table 9.4). There is a distinct break

Table 9.4. Measurements of Squash (*C. pepo*) Seeds from the Rhoads Site (11Lo8)

Fea. No.	Length (mm)	Width (mm)	Remarks
2	18.5	9.2	U.C.
40	18.4	10.1	
46–70	17.4	9.1	
40	17.4	8.3	
41	16.0	8.5	U.C.
118	15.6	?	U.C. broken
141	15.5	9.3	
40	15.4	9.5	
41	15.1	9.2	U.C.
46	15+	?	U.C. broken
40	15.0	9.0	
141	15.0	6.2	
141	12.3	6.4	
79	12.1	5.5	Cot.
141	11.9	5.5	Cot.
46–70	11.8	6.8	U.C.
40	11.7	5.4	
46–70	11.1	5.3	Cot.
17	11.0	8.4	C.S.
141	11.0	6.6	Cot.
2	11.0	6.3	U.C.
141	11.0	5.4	
118	11.0	?	Cot. broken
46	10.6	6.8	U.C.
17	10.5	7.0	C.S.
37	10.5	5.7	
163	10.3	7.7	U.C.
46	10.3	6.9	U.C.
118	10.1	?	U.C. broken
2	10.0	6.5	C.S.
2	10.0	6.5	C.S.
118	10.0	4.8	Cot.
141	9.6	6.5	Cot.
2	9.6	6.3	C.S.
118	9.3	4.7	Cot.
2	9.2	6.9	U.C.
118	9.2	3.9	Cot.
141	9+	6.5	Cot. broken
141	8.5	4.4	Cot.
41	8.4	4.3	
37	7.9	5.4	

Carbonized, unless otherwise noted. *U.C.* = Uncarbonized; *C.S.* = uncarbonized, copper stain; *Cot.* = cotyledon only.

in size between a length of about 12 and 15 millimeters. Those greater than 15 millimeters are seeds of winter squash, or small pumpkins, like the ones used for Jack-o'-lanterns at Halloween. Some of those in the upper range of 8 to 12 millimeters long could have been from the round, green-striped fruits resembling the "Mandan" variety of squash grown by the Indians of the Upper Missouri River. One of these is illustrated in Will and Hyde (1968:67). Most of the smaller seeds were probably from summer squash, which is often eaten before the fruit is fully mature, and of which there are many varieties. All of the squash from this site, as far as could be determined, appeared to be of one species, *Cucurbita pepo*.

Watermelon

Seeds of watermelon (*Citrullus lanatus*), a plant that originated in the Old World (Whitaker and Davis 1962:2), were found in five, or 9 percent, of the Class 1 pits excavated. The watermelon was brought to Florida by the Spanish sometime prior to A.D. 1576 (Connor 1925:159). It is not known when the Natchez Indians first began growing the plant, but by the time the French first made contact with them at the end of the seventeenth century, they had been growing it long enough to have named one of their months "The Month of the Watermelons" (Swanton 1911:103). Reasons for acceptance of this plant by the Indians are not hard to find. The refreshing sweetness of the flesh, particularly attractive to a people with few sweets in their diet, was probably the principal incentive to cultivation. Watermelons may be raised successfully by the same methods used for growing native squashes, and the plant does well in a long, hot, continental summer. Another probable factor in the acceptance of this foreign plant is the excellent keeping qualities of certain varieties when stored in the ground (Benson 1966:516).

Sizes of watermelon seeds found and comparisons with seeds from other historic and protohistoric sites are shown in Table 9.5. Although the seed from Feature 2 is uncarbonized, and there probably has been some shrinkage of the carbonized seeds from other pits, the differences in size and shape may indicate the presence of more than one variety of watermelon. This would not be surprising, for even in Marquette's day, Indians in Illinois were growing more than one kind of watermelon. The narrative "Marquette's First Voyage, 1673–1677" (Thwaites 1897–1900:127–29) has a passage that, translated from the French, reads: "They [the Illinois Indians] also sow beans and melons, which are excellent, especially those that have red seeds." Watermelon is the only species of melon that has reddish brown seeds. Furthermore, because some watermelons have black seeds, the inference is that more than one kind was grown. It is entirely possible that one or more kinds of watermelon had diffused up from Florida or Mexico, and that French missionaries could have brought seeds of

Table 9.5. Watermelon (*Citrullus lanatus*) Seeds from the Rhoads Site (11Lo8) Compared with Those from Other Historic and Protohistoric Sites

Site name	Rhoads	Coal Pit	King Hill	Zimmerman
Site number	11Lo8	23Ve4	23Bn1	11Ls13
Approximate period of occupation	A.D. 1760–1820	A.D. 1790–1820	ca. A.D. 1670–1720	A.D. 1673–1691
Historic tribe	Kickapoo	Little Osage	Kansa (?) (Late Oneota)	Historic Kaskaskia
No. of seeds	7	5	2	52
Sizes of seeds measured in mm	9.4 × 5.0	9.9 × 6.0	8.8 × 5.1	7.8 × 4.9
	9.5 × 4.7	10.1 × 5.6	9.4 × 5.3	9.0 × 5.6
	9.5 × 5.0	10.2 × 6.0		9.4 × 5.2
	12.3 × 8.2*	10.3 × 6.8		9.9 × 5.0
		11.0 × 5.6		9.9 × 5.2

Carbonized, unless otherwise noted.
*Not carbonized.

other varieties. At any rate, by the time the Rhoads site was first occupied, several kinds of watermelon were grown throughout the eastern part of the country, up into Canada, and in the Illinois country (Benson 1966:515, 516).

Wild Plant Foods

Remains of cultivated plants were found in a total of thirty-four, or 64 percent, of the Class 1 pits, but wild plant foods were found in only thirty, or 57 percent, of such pits. Fragile remains of beans and squash each occurred in Class 1 pits more frequently than any one wild plant food, except the nearly indestructible, carbonized stones of the wild plum (*Prunus* sp.). These matters indicate a possibility that wild plant foods were of less importance than cultivated ones to the inhabitants of this site.

Carbonized stones and fragments of stones, probably of the American wild plum (*Prunus americana*), were present in twenty-three, or 43 percent, of the Class 1 pits and in two of the Class 2 smudge pits. Gilmore (1919:87) had this to say about the use of wild plums: "The fruit was highly valued for food, being eaten fresh and raw, or cooked as a sauce. The plums were also dried for winter use. They were commonly pitted before drying, but the Pawnee say they often dried them without removing the pits." Although Gilmore was writing about the historic Indians along the Missouri River during the early part of the twentieth century, his remarks can be considered of general application to the eastern United States. Ethnographic references to a large number of areas by Yanovsky (1936:32) indicate that such practices were widespread. Gilmore's observation about plums being sometimes dried without removing the pits suggests that the presence of plum pits may not be a reliable indicator of time of occupation.

Hackberry seeds (*Celtis* sp.) were present in nine, or 17 percent, of the Class 1 pits and in one Class 2 pit. The frequency of their occurrence is probably due more to the indestructible nature of the mineralized seeds than to the importance of this fruit as a food resource. Yanovsky (1936:19) gives references to use by Indians on the Plains, in the Midwest, and in New York state. He said, "The fruit with seeds is pounded fine and used as a flavor for meat, or mixed with parched corn and fat." Such preparation would leave little evidence. Some fruits were also certainly eaten out of hand. Earlier, the possibility was mentioned that seeds were sometimes introduced by animals, or by falling directly from nearby trees.

Carbonized seeds of hawthorn (*Crataegus* sp.) were present in three Class 1 pits. On the eating of hawthorn fruits, John Josselyn wrote in 1672, "The haws . . . [are] very good to eat and not so astringent as the haws in England" (Fernald and Kinsey 1958:232). It apparently depends on which one of the many species one eats. Some are so bad that they were used by Indians only in time of shortage, but others were "eaten fresh," or the "fruit squeezed by hand, made into cakes and stored for the winter" (Yanovsky 1936:30, 31). As of now, we have seen hawthorn seeds from only one other archaeological site besides this one, that being a Mississippian settlement of about A.D. 1200–1300 in southeastern Missouri (Cutler and Blake 1976:49). Now that flotation or water screening is becoming general practice, such seeds may turn up more frequently.

Uncarbonized seeds of the American lotus (*Nelumbo lutea*) were found in two of the Class 1 pits. The seeds, which resemble small acorns, are very hard and decay resistant. Those from this site were found 2 feet and 2.5 feet below the surface. We have seen uncarbonized lotus seeds among the plant remains recovered from the Crawford Farm Sauk and Fox site (11Ri81) of about A.D. 1790–1810 and from the Little Osage Coal Pit site (23Ve4) of A.D. 1790–1820 (Cutler and Blake 1976:45). There is no reason to believe that any of these was intrusive.

Gilmore (1919:70) gives a good description of the uses of this plant by the Indians on the Missouri River: "The hard, nut-like seeds were cracked and freed of their shells and used with meat in making soup. The tubers, also, after being peeled, were cut up and cooked with meat, or with hominy." Yanovsky (1936:25) gives ethnographic references to the use of this plant for food from Connecticut and New York on the east, to Nebraska and the Dakotas on the west. The plant also ranges to the south from Florida to Texas. Fernald (1908:392), in the seventh edition of *Gray's New Manual of Botany,* in speaking of the present range of the plant, suggests that it may have been an Indian introduction in Massachusetts and Connecticut. Yarnell (1964:59) says of the plant, "perhaps introduced generally or locally by the Indians." A new colony could easily be started by dropping a few viable seeds in a favorable spot, such as a shallow pond, or a quiet pool in a stream.

Seeds of wild grape (Vitaceae) were found in two Class 1 pits, three in Feature 41 and two in 147. Such a sparse showing of small fruits, which each contain several seeds, would be surprising if the use of such plants had been extensive.

One carbonized fragment of black walnut shell (*Juglans nigra*) was found in Feature 75, a Class 1 pit, and one partially carbonized whole nut was found outside of any pit, in general excavation. The only other nut recovered was a fragment of a hazelnut (*Corylus americana*) in Feature 18, a Class 2 smudge pit. The scarcity of remains of nuts, which are so abundant on most prehistoric Midwestern sites, could mean that animal fat and protein were so plentiful, when firearms were used for hunting, that nuts were not worth the trouble of gathering. It could also mean that the site may have not been occupied in the later part of fall. This would not explain the scarcity of hazelnuts, which ripen in September and could be expected to be common on forest edges.

One carbonized stone of wild chokecherry (*Prunus virginiana*) was found in Feature 118. A single, broken, carbonized pawpaw (*Asimina triloba*) was found in Feature 163. Both of these features were Class 1 pits. We do not think that it is proper to try to draw conclusions from the single occurrence of a wild food, because of the many accidental ways that it could have become incorporated into the deposits. On sites where pawpaw seeds are abundant, as on the King Hill site (23Bn1) of about A.D. 1700 in St. Joseph, Missouri (Blake and Cutler 1982:101), they provide a good seasonal indicator. They ripen in that latitude about the early part of October. Ripening time is limited, and storage for any length of time is difficult, because of the perishable nature of the fruit. The seeds are easily identified, because of the striated nature of their inner structure, which is preserved even on broken and carbonized specimens.

The uncarbonized seeds of several different kinds of plants, other than those discussed above, were recovered by water screening the contents of pits. Some, and possibly all, of these may be intrusive. The black, round, shiny seeds of pokeweed (*Phytolacca americana*) were present in twenty-eight, or 53 percent, of the Class 1 pits excavated and in three of the Class 2 pits. It is well known that the first shoots of the plant were relished by the Indians, and still are today, especially in the Southern states. Fernald and Kinsey (1958:187, 188) wrote, "Some writers state that the berries are used for making pies and tarts. Others, however, state that the seeds are poisonous." We have frequently observed that the berries are relished by birds. We have no knowledge as to whether the seeds are resistant to decay, or whether they were blown into the open pits, or how they got there.

The case of small, light seeds of members of the blackberry and raspberry genus (*Rubus* sp.) is somewhat similar, except that we do have some evidence of their resistance to decay under favorable conditions. Seeds of *Rubus* sp. were present in eighteen, or 34 percent, of the Class 1 storage pits, later used for refuse (Table 9.6). No *Rubus* seeds were found in smudge pits.

In 1969 we identified masses of uncarbonized seeds of *Rubus* sp. from a level 24 to 30 inches above the bottom of a stone-walled privy, containing 7 feet of deposits, which was excavated by Robert T. Bray at historic Hanley House in Clayton, a suburb of St. Louis. Mr. Bray said, "It appears that the privy started receiving trash in small quantities almost from the time of construction (estimated at ca. A.D. 1855), but that no intensive filling occurred until the last decades of the 19th century. At that time (ca. A.D. 1880–1890) the 'fill' in the privy was surprisingly shallow—only about 24–30 inches had accumulated" (Bray 1971:6). Bray (personal communication) has estimated the age of the seeds at fifty to 100 years. Because the uncarbonized seeds, mixed with remains of pulp, were in masses 5 to 6 inches deep, it is reasonably certain that they neither blew in nor had fallen in accidentally. We have also seen uncarbonized seeds of *Rubus* sp. from the King Hill site (23Bn1) of about A.D. 1700 (Blake and Cutler 1982:102), although, as with those from the Rhoads site, we are not sure how the seeds got into the deposits. They were not present in all levels, however.

Table 9.6. Comparisons of Plant Remains from Subclasses of Class 1 Pits at the Rhoads Site

	Subclasses of Class 1 Pits					Percent of Class 1 Pits Excavated
	A	B	C	D	Total	
Class 1 pits excavated	19	22	11	1	53	100
Pits containing plant remains	17	20	9	1	47	89
Cultivated Plants						
Corn	12	18	3	—	33	62
Beans	8	5	—	—	13	25
Squash**	7	7	—	—	14	26
Watermelon**	3	2	—	—	5	9
Total cultivated plants	13	18	3	—	34	64
Wild Food Plants						
Wild plum (*Prunus* sp.)	9	8	6	—	23	43
Hackberry (*Celtis* sp.)***	6	3	—	—	9	17
Hawthorn (*Crataegus* sp.)	1	3	—	—	4	6
Lotus (*Nelumbo lutea*)*	1	1	—	—	2	4
Wild grape (?) (Vitaceae)	1	—	1	—	2	4
Black walnut (*Juglans nigra*)	—	1	—	—	1	2
Pawpaw (*Asimina triloba*)	—	1	—	—	1	2
Chokecherry (*Prunus virginiana*)	1	—	—	—	1	2
Total wild foods	14	10	6	—	30	57
Total wild and cultivated	16	18	7	—	41	77
Possibly Intrusive Plant Remains*						
Pokeweed (*Phytolacca americana*)	9	14	4	1	28	53
Blackberry (*Rubus* sp.)	5	7	5	1	18	34
Smartweed (*Polygonum* sp.)	3	2	1	—	6	11
Tick clover (*Desmodium* sp.)	—	2	—	—	2	4
Modern corn grains	—	2	1	—	3	6
Soy bean	—	—	—	—	1	2
Total possibly intrusive	9	15	5	1	30	57

Carbonized, unless otherwise noted.
*Uncarbonized.
**Some carbonized, some uncarbonized.
***Mineralized.

Uncarbonized seeds of smartweed (*Polygonum* sp.) were found in six Class 1 pits, tick clover (*Desmodium* sp.) was found in two pits, and a modern soybean (*Glycine max*) and a grain of modern dent corn were found in still another. Uncarbonized corn grains were also found in two other pits. All of these items appear to be modern intrusions, fallen from the edge of the pit, or brought in on the equipment or clothing of the excavators.

Observations

Remains of cultivated plants were found in thirty-four, or 64 percent, of the fifty-three Class 1 pits excavated. Except for Features 55 and 145, all such pits were in the western half of the site. Of the sixteen pits in the western half, north of Feature 105, that contained cultivated plant remains, eight contained beans, but only one contained watermelon seeds. Of the thirteen pits in the southern part of the western half, that is, south of Feature 100, five contained beans, and watermelon was found in four. Squash seeds were in eight of the pits in the northern group and in five of the southern. There were also differences in the proportions of row numbers among the corn cobs examined. Seventy-six percent of the twenty-nine cobs from the northern group were eight-rowed, 17 percent ten-rowed, and 7 percent twelve-rowed. One of the two twelve-rowed cobs (from Feature 40) was popcorn, the only example of this kind of corn that was seen from this site. Of the fifty-six cobs from the southern group, 89 percent were eight-rowed and 11 percent ten-rowed. No twelve-rowed cobs were seen.

These differences could be the result of chance, for the samples are small. They could be due to occupation in different years, to specialization by different parts of the population, or to other unknown factors (see Table 9.7 for a summary of the data discussed above).

Records for the first forty years of the twentieth century at the weather station in Lincoln, Illinois, near the Rhoads site, show the average date for the last killing frost in the spring was April 29, and the average date for the first in the fall was October 15, giving an average frost-free season of 169 days (U.S. Department of Agriculture 1941:842). Although the dates of killing frost might vary as much as a month each way in spring and fall, the usual variation was probably less than half as much, judging by the figures from other nearby weather stations (U.S. Department of Agriculture 1941:845, 846). Because successful corn planting usually takes place near the last expected frost in the spring, it appears probable that planting usually took place sometime in early May. Squash, beans, and watermelon are more sensitive to cold than corn. Planting of these crops could be expected to take place later in May, after the ground had become warmer.

Judging by these estimates of planting times, someone was probably on the site sometime in April for preparation of the ground, and through May and part of June for planting, cultivation, and hilling up of the corn. Green corn

Table 9.7. Summary of Cultivated Plants from Class 1 Pits on the Western Half of the Rhoads Site (11Lo8)

Feature No.	No. Corn Cobs by Row No.				Corn Grains	Bean	Squash	Watermelon
	8	10	12	Total				
North Pits								
1	N.M.				X	—	—	—
2	2	1	—	3	X	—	X	X
4	1	—	1	2	X	—	—	—
11	1	—	—	1	X	X	X	—
17	N.M.				—	X	X	—
22	1	1	—	2	—	—	—	—
37	2	—	—	2	X	—	X	—
54	—	—	—	0	X	X	—	—
75	5	1	—	6	X	X	—	—
40	1	—	1	2	—	X	X	—
46	3	1	—	4	X	X	X	—
47	1	—	—	1	—	—	X	—
48	1	1	—	2	X	—	—	—
46–70	3	—	—	3	X	X	X	—
71	1	—	—	1	X	X	—	—
74	—	—	—	0	X	—	—	—
Total	22	5	2	29	12	8	8	1
Percent	76	17	7	100				
South Pits								
79	3	—	—	3	—	—	X	—
105	—	—	—	0	X	—	—	—
114	8	1	—	9	—	—	—	—
125	N.M.				—	—	—	X
127	2	—	—	2	X	—	—	—
128	—	1	—	1	X	X	—	X
140	1	—	—	1	X	X	—	—

continued

Table 9.7. *Continued*

Feature No.	No. Corn Cobs by Row No.				Corn Grains	Bean	Squash	Watermelon
	8	10	12	Total				
152	1	—	—	1	X	X	X	X
116	5	1	—	6	X	—	—	—
113	4	—	—	4	X	X	X	—
119	12	1	—	13	X	—	—	—
141	12	2	—	14	X	X	X	X
163	2	—	—	2	X	—	X	—
Total	50	6	—	56	10	5	5	4
Percent	89	11	—	100				

N.M. = Present, but not measurable; *X* = present; *dash* = absent.

should have been ready for eating in August, as should summer squash and green beans. These crops would not be fully matured for the main harvest until September, at which time watermelons, which had begun to ripen in August, would be ready for picking. Of the wild foods, some plums and cherries would become ripe in July, with others not maturing until August. Hazelnuts would be ready to eat in September, with hawthorn fruits, lotus nutlets, and black walnuts. Different species of wild grapes would be ripening from summer to late fall. In this latitude, pawpaws should be ripe in early October (see Table 9.8 for summary of planting and harvest times).

People may not have been present in numbers during the summer, although a few were probably there in July to protect the crops from birds and animals and to gather wild plums. Harvest and processing of crops almost certainly kept many women and children, with men to protect them, on the site from August into early October. People could have been present at times other than those imposed by the planting and harvesting of crops, however, and the many storage pits suggest the possibility of a longer residency.

Dr. Walter Klippel, who directed excavation of the site, in his unpublished report (Sec. 4, Archaeology—Features) describes three subclasses of storage pits that are included under his Class 1 type. These are Subclass A, a conoidal pit with an average volume of about 42 bushels; Subclass B, a cylindrical pit with a very slightly smaller than average volume; and Subclass C, a basin-shaped pit with an average capacity of approximately 29 bushels. It is not known whether differences among the three subclasses represented

Table 9.8. Approximate Harvest Times of Cultivated and Wild Plants

	Approximate Planting Time	Approximate Harvest Time
Cultivated Plants		
Corn	May	Late August, September
Green Corn	May	Early August
Beans	Last half of May	August, September
Squash	Last half of May	August, September
Watermelon	Last half of May	Late August, September
Wild Plants		
Wild plum (*Prunus* sp.)		July, August
Hackberry (*Celtis* sp.)		September, October
Hawthorn (*Crataegus* sp.)		September, October
Lotus (*Nelumbo lutea*)		September, October
Wild grape (Vitaceae)		September, October
Black walnut (*Juglans nigra*)		September, October
Pawpaw (*Asimina triloba*)		September, October
Chokecherry (*Prunus virginiana*)		July, August
Hazelnut (*Corylus* sp.)		September

functional or cultural differences, or whether the differences were dependent on the personal preference or diligence of the individual or individuals who dug them.

Table 9.6 was drawn up in an effort to analyze by subclasses the contents of Class 1 pits that contained food plants. It is realized that carbonized contents could not give a clue as to what was originally stored in a pit, for carbonized food that was inedible would be more likely to turn up in an abandoned pit than in one still in use. Table 9.6 does indicate some differences, but we are uncertain whether they are meaningful. Subclass B pits contained the largest proportion of cultivated plant remains, Subclass A was next, and Subclass C was last. Out of eighteen Subclass B pits that contained wild and cultivated plant remains, eighteen contained cultivated and ten wild. Out of sixteen Subclass A pits, thirteen contained cultivated and fourteen wild. Out of seven Subclass C pits that contained wild and cultivated plant foods, three contained only one cultivated plant, corn, and six contained evidence of wild food, mostly wild plum.

1. Since this paper was written, there has been a reexamination of the plant remains from under Mound 51. In the light of this later study, it now seems possible that some of the larger seeds are those of the plant *Cucurbita argyrosperma* ssp. *argyrosperma*, because of the recovery of an uncarbonized peduncle of *C. argyrosperma* in the later study. Seeds of *C. pepo* and those of *C. argyrosperma* may be difficult to tell apart (Fritz 1994).

10 Plants from Archaeological Sites East of the Rockies

Hugh C. Cutler and Leonard W. Blake

Missouri Botanical Garden

(A text version of a 1976 microfiche, published by the Missouri Archaeological Society, University of Missouri, Columbia, which was a corrected and updated version of the 1973 edition)

Leonard Blake's Comments, 1999

In 1973, a report on identification of plant remains that had been sent to Cutler and Blake at the Missouri Botanical Garden was hastily produced in an edition of 200 mimeographed copies. The purpose was to let those who had sent us material know what had been done with it. It proved to be so popular that we ran out of copies to distribute.

In 1976, we were provided with the opportunity to correct mistakes and bring the report up to date by means of a microfiche edition, which was produced through the Department of Anthropology at the University of Missouri-Columbia. A change in administration brought about a cancellation of this program soon after it started, so distribution was very limited. We are now putting out the contents of this 1976 edition in printed form.

Information in this chapter was assembled over a period of years when important archaeological excavations were being carried out on the Plains, and in the rest of the eastern United States, yet few others were attempting to collect, identify, record, analyze, and publish information on corn and other plant remains. It seems reasonable to expect that this chapter has a value for these reasons, even though the samples received were usually unsystematically collected "grab" samples, and identifications and measurements were usually made without use of a microscope. Analyses (even if sometimes primitive by today's standards) would, otherwise, not have been made.

In hindsight, it would have been good if there had been more measurements on corn, if magnification had been used more, if more attention had been paid to wild plants, and if Cutler and I had worked on this "full time." Cutler was acting director of the Missouri Botanical Garden at that time, and I came to the garden only two days a week.

Our files on material covered in this report are included in those stored at the Illinois State Museum, Springfield, under the name of "Cutler Blake." Most include additional information, and most of those on corn have coordinated graphs like those used in Chapter 4 of this volume.

"Plants from Archaeological Sites East of the Rockies" was not continued beyond 1976, mainly because of Cutler's retirement in 1977, and because at that time archaeologists were beginning to systematically retrieve and study plants, usually using some form of flotation or water screening.

NOTE: Comments about Middle Woodland maize are now outdated, and shown to have been based on false assumptions by recent accelerator radiocarbon dates.

Cultures and dates shown in this report are those received with the material analyzed. Since 1984, accelerator radiocarbon dates on maize have been published from three sites, which have shown that the maize from these is not so old as other material from the same feature.

One site is Jasper Newman, in Moultrie County, Illinois, where a carbon 14 date of A.D. 90± was obtained on charcoal, but where maize from the same pit was dated at only 450 ± 500 years B.P. (Conard et al. 1984).

The other two sites are the McGraw site in Ross County, Ohio, and the Daines II Mound in Athens County, Ohio. McGraw had bone dated at A.D. 230 ± 80 years, and Daines II at 280 ± 140 B.C., but maize from each of these sites had an accelerator carbon 14 date of less than 400 years B.P.

The plants people use are a key to the past and to the activities and environment at the time the plants were collected. The assemblage of plants, wild and cultivated, used in any community is unique and can be used to trace the history of the people. How much we discover of the relationships of human groups with their environments, and their neighbors, and the parallel evolution of plants and culture depends upon the number and depth of collections and studies available from a region, from related areas, and from other time periods, and upon how well we understand each plant.

Methods for the recovery and study of plant and animal remains are still relatively crude. The century-old techniques of water screening and flotation as tools for recovery of plant and animal remains are now frequently used but more people should be trained to save plant materials and to study and interpret the results in terms of plants, environment, and people. There must be more effective communication and interaction between biologists and students of humankind.

We pay most attention to maize because so much of it is recovered and there are enough characters in even fragmentary parts of cobs so that differences and similarities can be measured, and valid comparisons made between collections. Cultivated plants may be considered artifacts, the result of human activities and dependent upon people for their survival as well as their development. But crop plants are also affected by factors other than cultural ones. The crop a farmer harvests is always different from the one harvested the year before, and from the seed originally planted.

Summaries of our viewpoints may be read in Cutler (1966, 1968), Cutler and Blake (1971), Cutler and Meyer (1965), Cutler and Whitaker (1967), and Whitaker and Cutler (1971).

Most American Indians grew several very different kinds of corn. They usually selected seed to definite standards and planted the kinds in separate fields and at different times. We can make useful comparisons from fragments of cobs or a few grains, but many kinds of corn can be distinguished only when we have entire ears. In the table we have lumped all corn from a site or from a recognizable period.

The single most useful character for study of the corn ear is the number of rows of grains. It may vary slightly under different growing conditions and with different positions on the plant, but row number is relatively constant for a recognizable kind of corn. It can often be estimated from a single grain or from a fragment of an ear. Like the sides of a piece of pie cut into eight parts, the sides of a grain from an eight-rowed ear make an angle of 45 degrees.

A pair of grains is borne in a cupule, the basic unit of an ear. The sides of a cupule of an eight-rowed ear will thus form an angle of 90 degrees. The angle formed by the sides of a cupule from a ten-rowed ear will make an angle of 72 degrees (two grains, or rows, borne in each of five cupules equals one-fifth of 360 degrees). The number of rows of grains can sometimes be determined from a single cupule.

Cupule width, the distance across the entire pocket in which a pair of grains (kernels) and their spikelets is borne is a measure of the size of the cob because the points of measurement can be determined accurately and because it is less affected by the number of rows of grains. Most corn cobs shrink from 15 to 20 percent when they are carbonized so an adjustment must be made when carbonized and uncarbonized specimens are compared. Nearly all collections in this list were carbonized and no adjustment has been made.

We depend upon the data sent with collections and on packing lists. When we receive bags identified as coming from a definite area, with a map of the site and indications of relationships and of relative ages, and indications that the material is from a burial, a cache, a certain area in a burned house or room, or from general rubbish of a certain area of a site, we can make useful comparisons. We can see what came from different parts of the site; what kinds of materials were early or late. It helps to know how a specimen was deposited, as well as where. This is especially important with wild plant remains because they may have been introduced into the site without human intervention before it was built, during occupation, or after it was abandoned; or the material may have been introduced accidentally or purposefully by human action.

Most of our time is spent on cultivated plants because so many collections come in that we have little time for the wild ones. It would be best if we could do our own sorting of the cultivated plants and the related wild ones from the mass of plant materials because many fragments can be identified only after much experience, but we can do this only when collections are small and we have help. Occasionally we have identified wild plants but our notes in this list are erratic and should not be used for quantitative comparisons. Figures for amounts of plant materials from archaeological sites are unreliable because there are so many variables and accidents. Selective deposition, preservation, recovery, submission for study, and even the depth of the study make results comparable only with caution.

Corn apparently originated in western central Mexico where there is great diversity in kinds grown today and where the largest number of its close relatives, teosinte (*Euchlaena* spp.) and *Tripsacum* spp., grow. Some of the early corn found in archaeological sites in the United States is eight-rowed, but the earliest corn from Bat Cave (Dick 1965), which is a few miles east of the continental divide in New Mexico, and Tularosa Cave (Cutler 1952), which lies a few miles west, is mainly twelve-rowed. In Tularosa Cave levels after A.D. 700, the majority of cobs have eight rows of grains.

In the central and eastern states, most of the earliest corn has small cobs and cupules, and about twelve rows of grains. We can call this race North American Pop. It has evolved from early, hard, and small southern corn and has sometimes been called Tropical Flint or Reventador. Some of this ancient kind of corn is still grown by a few modern Indians.

Eastern Eight Row, first described as Northern Flint, derived from eight-rowed forms of the Southwest and Mexico. It reached Ontario, Canada, before A.D. 800. By A.D. 1200, Eastern Eight Row dominated most of the region east of the Mississippi and by A.D. 1500 covered most of the region to the Rockies. Today Eastern Eight Row continues to be important because crosses of it with dent corn produced the Corn Belt Dent upon which most commercial corn production in North America is based.

Intermediate to North American Pop and Eastern Eight Row is Midwest Twelve Row. It is similar to the Pima-Papago corn race of the Southwest and persisted longest in parts of the central and lower Mississippi River Valley. By A.D. 1500 it had largely disappeared but left its mark in the common ten- and twelve-rowed variants of Eastern Eight Row that are grown in the Plains area, especially towards the south.

A remarkable feature is the scarcity of corn until A.D. 900 or 1000. This could indicate gaps in collections or an actual scarcity of corn agriculture during this period. Descendants of the early small-cobbed and many-rowed corn are still being grown in many places as hard flints or popcorns, but by about A.D. 1200 most corn east of the Rockies was eight-rowed.

We have no reliable record of pre-Columbian dent corn in the eastern United States. While dent corn had been grown for a short period by Fremont culture farmers in Utah and by some Pueblo Indians before A.D. 1250, extreme forms of dent corn appear to have been introduced in the Rio Grande valley by Spanish colonists about A.D. 1700. On their return after the Pueblo Rebellion, dent corn probably was grown in many southern states in the 1700s, but our only specimens are from a Spanish mission site in Texas and from the Nuyaka site in Alabama. An English clergyman, John Banister, mentions dent corn in Virginia in a letter written in April 1679 (Ewan and Ewan 1970). Present-day commercial field corn is a less than 200-year-old hybrid of the eight-rowed Eastern Eight Row with the European-introduced, many-rowed Mexican dents (Anderson and Brown 1952).

All of the squash and pumpkins we have seen from pre-Columbian sites east of the Rockies have been *Cucurbita pepo,* the species to which belong the Halloween pumpkin, acorn, summer crookneck, zucchini, white bush scallop, and similar squashes and the small, yellow-flowered ornamental gourd. The last is usually considered a distinct variety, *C. pepo* var. *ovifera.* Several Gulf Coast sites have yielded small seeds that belong to this variety. The early thick-shelled small squashes from Salts Cave, Kentucky, are developments of the small-seeded gourds and are related to small eastern squashes like the summer squash.

The cushaw (*C. mixta*) might be found in sites just east of the Rockies after about A.D. 1000, and a squash common in warm regions, *C. moschata,* is to be expected in sites just north of the Mexican border. Both these species have been found in sites to the west but not east of the Rockies. Hubbard, Mexican banana, turban, and similar types of the South American squash, *C. maxima,* apparently were not important in our area until the late 1700s. Seeds similar to those of the hubbard squash were recovered from Fort Berthold on the Missouri River and are dated about 1845–62.

The oldest cultivated plant may be the bottle gourd (*Lagenaria siceraria*). It does not tolerate cold weather so, near the northern margins of its range and northwards, its role as a container is sometimes taken over by thick-walled forms of *Cucurbita pepo.* Rinds of the bottle gourd and of thick-walled forms of *Cucurbita pepo* have been recovered from Salts Cave, Kentucky, and from Sheep Rock Shelter, Pennsylvania.

ALABAMA

Site Name & No.	Culture/Tribe	Date	Location	Sample furnished by	Mean Row No.	Median Cupule Width (mm)	No. Cobs	Row Numbers % of Total Cobs					Other Plant Remains/Comments
								8	10	12	14	16+	
Nuyaka Village	Historic Upper Creek	AD 1777-1813	Horseshoe Bend Natl. Park, Tallapoosa Co.	Chas. Fairbanks	12.4	8.2	28	4	21	39	25	11	
Fort Toulouse (1EL8)	French Trading Post to Creeks	AD 1717-1776, perhaps early 1800s		D. L. Heldman	8.0	7.8	8	100	—	—	—	—	Peach
Talisi (1MC1)	Historic Creek	AD 1710-1750, perhaps to 1840	Tulsa	J. W. Cottier	9.2	9.0	17	41	59	—	—	—	Acorns
Town of Taitt (1RU63)	Historic Yuchi	AD 1685-1836	Walter F. George Res.	R. L. Stephenson	9.0	9.0	52	60	34	2	4	—	22 Small Peach Pits; Black Walnut
Town of Taitt (1RU63)	Historic Yuchi	AD 1685-1836	Walter F. George Res.	H. A. Huscher	9.4	8.3	168	47	38	12	3	—	Measurements on 39 cobs; All but 31 of 168 cobs are from small, corn cob filled pits
1BR46	Late Creek		Walter F. George Res.	H. A. Huscher	11.4	7.6	25	4	60	8	20	8	
1HE34	Fort Walton-Lamar		Walter F. George Res.	H. A. Huscher	8.5	5.8	4	75	25	—	—	—	Hickory Nut, Acorn
Seaborn Mound (1HO27)	Early Fort Walton	AD 1500-1600	Columbia Dam & Lock	R. W. Neuman	8.4	8.0	5	80	20	—	—	—	"Probably Eastern Flint"
1BR21	Weeden Island		Walter F. George Res.	H. A. Huscher									Hickory Nut, Black Walnut
1HO24	Weeden Island	ca. AD 1000-1300	Columbia Dam & Lock	R. L. Stephenson									6 Tubers Groundnut (<u>Apios americana</u>)
1RU58	Late Swift Creek-Weeden Island Affiliation	ca. AD 700-1000	Walter F. George Res.	H. A. Huscher									Hickory Nut
1EL52	Autauga Phase Late Woodland	C^{14} AD 920		D.W. Chase									Hickory Nut, Persimmon, Acorn, Hackberry, Hazelnut
1EL52	Hope Hull Focus, Late Middle Woodland underlying Late Woodland	Late Woodland date of C^{14} AD 920		D.W. Chase	9.5	7.2	13	38	54	—	8	—	"Eastern Flint - Flour Mainly"
1BR35	Culture?	Date?		J. Cottier	10.0	7.3	6	33	33	33	—	—	8-rowed cobs have closed cupules. Those of the 10 and 12-rowed are more open.

ARKANSAS

Site Name & No.	Culture/Tribe	Date	Location	Sample furnished by	Mean Row No.	Median Cupule Width (mm)	No. Cobs	Row Numbers % of Total Cobs						Other Plant Remains/Comments
								8	10	12	14	16+		
Adair (3GA1)	Caddo Late	ca. AD 1400+		J. C. Weber	11.0	6.0	77	12	39	39	10	—	Common beans, peach
Albertson Shelter #1 (3BE174)	Neosho Focus?	Recent?		D. Dickson	14.0	7.8	2	—	—	—	100	—	Not carbonized
Shawnee Village	Mississippian, Nodena Phase	ca. AD 1400-1700	Mississippi County	G. Perino	10.0	7.0	2	—	100	—	—	—	
Upper Nodena (3MS4)	Mississippian, Nodena Phase	ca. AD 1400-1700		D. F. Morse	10.9	6.9	97	12	41	38	9	—	Hickory, black walnut, persimmon, pawpaw, pecan, wild cherry
Parkin Village (3CS29)	Mississippian, Parkin Phase	ca. AD 1400-1700		G. Perino	10.0	8.3	2	—	100	—	—	—	Hickory nut, pawpaw
Barton Ranch (3CT18)	Mississippian, Parkin Phase	ca. AD 1400-1700		G. Perino	10.1	7.3	20	30	35	35	—	—	
Kappenberger (3MS3)	Late Mississippian			T. C. Klinger	10.0	7.3	1	—	100	—	—	—	Hickory nut
Moore Bayou (3AR14)	Late Mississippian	ca. AD 1400-1600		B. B. McClurkan	11.2	5.8	5	—	40	60	—	—	
Lawhorn	Middle Mississippian	ca. AD 1550	Craighead County	R. Marshall	11.4	6.8	7	—	29	71	—	—	
Banks	Middle Mississippian	C¹⁴ on corn AD 1535 ± 150	Crittenden County	G. Perino	11.0	5.4	51	4	47	43	6	—	Persimmon, also corn grains from 8, 10, and 12-rowed ears.
Vernon Paul (3CS25)	Mississippian			D. F. Morse	10.7	6.4	3	34	33	—	33	—	
Charles MacDuffie	Middle Mississippian	(Earlier than Banks Site)	Craighead County	F. J. Soday	12.0	6.3	26	4	23	46	23	4	
Lester Place	Caddo	ca. AD 1200-1400	Lafayette County	G. Perino (Lemley Collection)	13.7	5.6	6	—	—	33	50	17	"12-14-rowed moderately deep, thick-walled cupules." "16-rowed deep cupules, thin-walled." "Hard flint or pop corn."
Hays Mound (3CL6)	Caddo	ca. AD 1100-1300		J. C. Weber	12.0	5.3	3	—	33	33	33	—	Also corn grains from 8, 10, & 14-rowed ears 4.9 to 9.2 mm wide. Hickory nut, black walnut, persimmon, acorn, pecan, Ampelopsis, Lithospermum, Paspalum sp., Desmodium
Zebree (3MS20)	Early Mississippian, Big Lake Phase	ca. AD 900-1100		D. F. Morse	11.8	5.7	18	11	22	34	33	—	Hickory nut, Hackberry, Lithospermum sp.
Derossitt (3SF49)	Baytown - Mississippian			Carol Spears									5 fragmentary corn grains carbonized off the cob, 5.3 to 8.5 mm wide; persimmon, wild grape

Archaeological Sites East of the Rockies 99

ARKANSAS

Site Name & No.	Culture/Tribe	Date	Location	Sample furnished by	Mean Row No.	Median Cupule Width (mm)	No. Cobs	8	10	12	14	16+	Other Plant Remains/Comments
Breckinridge Shelter	Bluff Dweller? Late Woodland or Mississippian	ca. AD 800-1500 est.	Carroll County	Excavated by Harrington 1922; R. A. Thomas	10.3	8.0	20	25	40	30	5	—	Not carbonized; warty squash rind, small <u>C. pepo</u> peduncle, gourd rinds 2.6-5.0 mm thick
3NW31	Late or Transitional Woodland	ca. AD 1000+		D. R. Dickson	10.2	7.0	28	28	43	21	4	—	Not carbonized. Squash (<u>C. pepo</u>), gourd, common bean, hickory nut, black walnut, persimmon, acorn, hackberry, pawpaw, <u>Chenopodium</u>, <u>Amaranthus</u>, sunflower, wild grape, large <u>Iva</u>, honey locust, ragweed, wild bean (<u>P. polystachyus</u>), canary grass (<u>Phalaris</u> sp.), cocklebur (<u>Xanthium</u> sp.), cane, reed, milkweed, <u>Apocynum</u>.
War Eagle Creek Bluff Shelter	Culture?	Time?	Benton County	C. Perino	11.8	7.6	8	—	25	63	12	—	<u>Not carbonized.</u>
Butler's Ford Bluff Shelter	Culture?	Time?	Near Springdale	G. Perino	10.0	7.7	3	33	33	33	—	—	<u>Not carbonized.</u>
Shirley Cave Site	Culture?	Time?	Van Buren County	Don E. Crouch	12.0	7.0	1	—	—	100	—	—	<u>Not carbonized.</u>
Hidden Valley Shelter	Culture?	Time?	Newton County	G. Perino	9.0	8.6	2	50	50	—	—	—	<u>Not carbonized.</u>
Lost Valley Shelter	Culture?	Time?	Newton County	G. Perino	10.7	8.0	3	33	67	—	—	—	<u>Not carbonized.</u>
Indian Creek Bluff Shelter	Culture?	Time?		G. Perino	10.0	9.1	1	—	100	—	—	—	<u>Not carbonized.</u>
3LO4	Culture?	Time?	Dardanelle Reservoir	R. L. Stephenson									Black Walnut, persimmon, acorn
3PP1	Culture?	Time?	Dardanelle Reservoir	R. L. Stephenson									Hickory nut (<u>C. ovata</u>)
3YE4	Culture?	Time?	Dardanelle Reservoir	R. L. Stephenson									Hickory nut (<u>C. ovata</u>)

FLORIDA

Site Name & No.	Culture/Tribe	Date	Location	Sample furnished by	Mean Row No.	Median Cupule Width (mm)	No. Cobs	Row Numbers % of Total Cobs					Other Plant Remains/Comments
								8	10	12	14	16+	
8JE1	Spanish Mission	AD 1675-1704	16 mi. SE of Tallahassee	H. McAleenan	8.0		5	100	—	—	—	—	Also distorted corn grains, mostly 8-row and at least one 10-row. "Cupules open to moderately longitudinally compressed." 60 common beans, acorn, honey locust, cane
Velda (8LE44)	Late Ft. Walton - Early Appalachee	Should be early 17th Century		L. R. Morrell	8.0	6.7	1	100	—	—	—	—	Cupule more open than on typical northern flint. One common bean, 11.0 mm long. 5.8 mm wide
8JE5	Ft. Walton	ca. AD 1400	Jim Woodruff Reservoir	R. Bullen									Corn grains mostly 8- and 10-rowed, some 12-rowed. Eight rowed are crescent shaped.
Curlee (8JA185)	Early Ft. Walton &/or Late Weeden Island			P. S. Essenpreis	8.6	8.0	24	75	21	4	—	—	Acorns
Patton Seslie Site (8WA39)	Culture?	ca. AD 1000-1300		E. V. Komarek, Sr.	10.6	8.1	55	15	53	23	7	2	
Cayson (8CA3)	Culture?	ca. AD 1100-1200		P. S. Essenpreis									3 corn grains, charred off the cob, 8.2 to 9.0 mm wide.
Key Marco	Prehistoric	Date?	Collier County	R. A. Yarnell									Recovered by Cushing in 19th Century. Small forms of Lagenaria and seeds C. pepo var. ovifera.

GEORGIA

Site Name & No.	Culture/Tribe	Date	Location	Sample furnished by	Mean Row No.	Median Cupule Width (mm)	No. Cobs	Row Numbers % of Total Cobs					Other Plant Remains/Comments
								8	10	12	14	16+	
9LB8	Historic Spanish Mission	AD 1590-1670		D. F. Morse via J. B. Griffin	8.8	7.5	20	60	40	—	—	—	
9CLA41	Ft. Walton-Lamar	ca. AD 1200-1500		R. L. Stephenson	10.0	6.0	1	—	100	—	—	—	8 tubers groundnut (<u>Apios americana</u>)
Etowah Mound "C"	Last half Mississippian Period	ca. AD 1050-1400	Bartow County	L. H. Larson, Jr.	8.4	7.3	64	78	22	—	—	—	Also crescent-shaped corn grains. Possible fragment common bean, hickory nut, wild plum, acorn, <u>Viburnum</u> sp.
9QU1	Lamar and later overlying Archaic		Walter George Res.	H. A. Huscher and R. L. Stephenson	10.0	7.4	16	31	38	31	—	—	4 1/2 quarts charred, shelled acorns.
9ER54	Early Lamar & Weeden Island	ca. AD 1000-1300	Columbia Lock and Dam	R. L. Stephenson	14.0	8.2	1	—	—	—	100	—	3 tubers groundnut (<u>Apios americana</u>)
9CLA15	Lamar back thru Weeden Island	ca. AD 800-1400	Walter George Res.	R. L. Stephenson	12.0	6.4	2	—	—	100	—	—	
Lingerfeld (9WD1)	Late Woodland	ca. AD 900-1100		D. F. Morse via J. B. Griffin	9.7	8.9	121	37	43	20	—	—	
Williams (9GO507)	Early Woodland	before AD 1		D. F. Morse via J. B. Griffin	10.1	7.5	28	18	57	25	—	—	Corn said to be in Early Woodland association, but C14 date on charred wood was AD 1480±75 (M-1107)
9CE66	Culture?	Time?	Walter George Res.	H. A. Huscher	9.4		43	42	49	9			
9CLA52	Culture?	Time?	Walter George Res.	H. A. Huscher									Hickory nut fragments
9ME60	Culture?	Time?	Walter George Res.	H. A. Huscher									Hickory nut, black walnut, persimmon
9QU4	Culture?	Time?	Walter George Res.	H. A. Huscher									Hickory nut, pecan
9QU5	Culture?	Time?	Walter George Res.	H. A. Huscher									Hickory nut

ILLINOIS

Site Name & No.	Culture/Tribe	Date	Location	Sample furnished by	Mean Row No.	Median Cupule Width (mm)	No. Cobs	8	10	12	14	16+	Other Plant Remains/Comments
Crawford Farm (11RI81)	Historic Sauk and Fox	ca. AD 1790-1810		E. Herold	8.3	8.4	1300	86	12	2	—	—	C. pepo, butternut, black walnut, lotus, blackberry, Scirpus sp.
Rhoads (11LO8)	Historic Kickapoo	ca. AD 1760-1820 (probably 1812)		W. Klippel	8.4	8.3	262	85	12	3	—	—	One example of popcorn; common beans, summer squash and pumpkin types of C. pepo, watermelon seeds of 2 sizes, wild plum, hackberry, hawthorn, lotus, wild grape, pawpaw, black walnut, hazelnut, wild cherry, Rubus sp.
Waterman (11R122)	Historic Michigamea	ca. AD 1750-1770		M. K. Brown	9.5	7.5	421	40	47	12	1	—	
Knoll Spring Site (11CK19)	Sagaunashke Complex, Blue Island, Upper Mississippian	ca. AD 1750		C. M. Slaymaker, III									Northern flint corn grains, distorted, 9 to 12 mm wide, 6.6 to 8.1 mm long; 2 common beans, hickory and black walnut, leaf bases of rush, perhaps Carex sp.
South Terrace, Monks Mound (11MS38)	Possibly Historic	ca. AD 1700+	Madison County	C. J. Bareis	8.9	7.9	40	65	28	7	—	—	
Palos (11CK26)	Blue Island (Huben) Focus, Upper Mississippian	ca. AD 1675-1700		P. J. Munson & M. K. Brown	8.9	8.2	21	71	19	10	—	—	Also crescent-shaped corn grains; common beans, sunflower, hickory nut, wild plum, black walnut, acorn, hazelnut, Chenopodium or Amaranthus sp.
Zimmerman (11LS13)	Historic Kaskaskia	AD 1673-1691		E. Herold & M. K. Brown	8.6	8.0	22	68	32	—	—	—	Common beans, watermelon, hickory nut, wild plum, hackberry, pecan, hazelnut, wild cherry, wild grape, black haw (Viburnum), hawthorn
Plum Island (11LS2)	Proto-historic Kaskaskia	ca. AD 1500-1600	LaSalle County	E. Herold	9.7	7.4	17	38	41	20	1	—	+59 corn grains.
Hoxie Farm	Upper Mississippian	ca. AD 1550	Near Thornton, Cook County	D. Pedric & E. Herold	9.8		10	40	30	30	—	—	Corn grains
Stolle Quarry	Upper Mississippian, Oneota	est. AD 1400	St. Clair County	G. Perino	8 or 10	6.0	1 fragments						
Crable	Middle Mississippian	C14 AD 1340-1430±100		Letter R. A. Yarnell	11.0	7.1	16						
Merrell Tract (11MS2-3)	Moorehead-Sand Prairie phase Mississippian	ca. AD 1250-1500		R. Salzer, M. Woodworth	9.4	6.1	118	45	40	14	1	—	Black walnut

ILLINOIS

Site Name & No.	Culture/Tribe	Date	Location	Sample furnished by	Mean Row No.	Median Cupule Width (mm)	No. Cobs	Row Numbers % of Total Cobs					Other Plant Remains/Comments
								8	10	12	14	16+	
East Ramp, Monks Mound (11MS38)	Sand Prairie phase, Mississippian	C¹⁴ AD 1310 ± 55		K. Williams	9.5	6.8	4	25	75	—	—	—	
Jasper Newman (11KS4), F-21, F-37	Middle Mississippian	C¹⁴ AD 1380 ± 100	Moultrie County	W. M. Gardner	10.5	7.6	13	23	53	8	8	8	Hickory, sunflower
Larson (11Fv1109)	Mississippian	ca. AD 1300±		P. J. Munson & A. Ham	10.0	6.7	17	35	35	24	6	—	common bean, hickory nut, wild plum
Hungry Wolf Site	Larson phase (?), Mississippian		Fulton County	L. Conrad	12.5	6.8	4	—	—	75	25	—	
Cedar Row Location, Dickson Mounds	Middle Mississippian (Plains Traits in Ceramics)	ca AD 1300± 100	Fulton County	J. Caldwell, P. J. Munson	9.5	7.2	14	29	50	21	—	—	
Emmons	Middle Mississippian	ca AD 1200-1400	Fulton County	Dr. D. Morse									110 corn grains charred and distorted, mostly 8-rowed.
11Fv47	Bold Counselor phase, Oneota	ca. AD 1300		Charles W. Cooper, L. A. Conrad	10.0	9.5	2	—	100	—	—	—	
Olin (11MSv279)	Late Woodland & Mississippian	ca. AD 1140-1400		S. Denny	10.1	6	52	29	40	27	4	—	Common bean, hickory nut, wild plum, black walnut, persimmon, acorn, wild grape. Rubus sp.
Kincaid	Middle Mississippian	ca. AD 1285 ± 75	Pope & Massac counties	R. A. Yarnell Letter 6/28/65	10.8	7.3	13	—	—	100	—	—	Cupules are open and the glumes moderately thick.
11Ov117	Mississippian	ca. AD 1100-1300		W. Klippel	12.0	6.8	1	—	—	16	—	—	
Orendorf (11Fv1284)	Early and Later Larson phase, Mississippian	C¹⁴ AD 1170±70 to 1250±65		L. A. Conrad	10.1	6.8	50	32	42	16	10	—	Also numerous corn grains. Hickory nut
First Terrace, Monks Mound (11MS38)	Mississippian, Moorehead phase	C¹⁴ 1110-1280±55		E. Benchley	12.0	8.0	1	—	—	100	—	—	
Angelly (11MX66)	Mississippian	ca. AD 1255+		B. M. Butler	10.8	7.2	5	—	60	40	—	—	Also medium corn grains, most 7-8 mm wide. Common bean, hickory nut, wild plum (P. hortulana & P. angustifolia), black walnut, persimmon, Viburnum, wild crabapple

ILLINOIS

Site Name & No.	Culture/Tribe	Date	Location	Sample furnished by	Mean Row No.	Median Cupule Width (mm)	No. Cobs	Row Numbers % of Total Cobs					Other Plant Remains/Comments
								8	10	12	14	16+	
Collinsville Airport (11S34)	Middle Mississippian	ca. AD 1085-1435 (C14)	St. Clair County	C. Bareis	11.3	7.0	19	11	26	53	10	—	1 10-rowed, possibly popcorn
Top, Monks Mound (11MS38)	Middle Mississippian	AD 1110±150	Madison County	C. J. Bareis	10.4	5.2	10	20	40	40	—	—	Also corn grains, mostly 8 to 10-rowed, small to medium
Under Mound 34	Middle Mississippian	AD 1150±100 (C14)	Madison County	G. Perino	11.9	6.4	27	7	19	48	22	4	Also large, medium, and small corn grains. Cucurbit sp. (pepo?), pawpaw, pecan, Scirpus vallidus, Eleocharis sp. stem, acorn
Hood (11Mv56)	Late Woodland	C14 AD 1000±100, AD 1230±115		R. B. Lewis									Small corn grains and fragments; hickory nut, black walnut, hackberry, possible pecan, ground nut (Apios)
Eveland II, Dickson Mound (11Fv900)	Mississippian	ca. AD 1144 (Aver. 3 C14 dates)		G. Schroeder									(CN 11-6) 1 carbonized ear, 10-rowed, C.W. 5.5 mm, (CN 11-2) fragments, cupules and grains, 10 12-rowed, C.W. 5.0 to 5.5 mm, small glumes, cob not thickened
Mound 19	Mississippian	ca. AD 1150 est.	Madison County	G. Perino	11.1	7.5	9	22	22	34	22	—	Corn grains, gourd rind, hickory nut, pawpaw, Chenopodium sp., grass seeds, Carex sp.
Loyd (11MS74)	Late Woodland & Middle Mississippian	est. AD 1150-1250	Madison County	R. Hall	11.1	6.7	63	10	38	43	8	1	Corn grains, persimmon
Under Mound 31	Middle Mississippian	ca. AD 1100-1200 est.	Madison County	J. R. Caldwell	12.0	6.4	1	—	—	100	—	—	1 8-rowed, 1 12-rowed corn grains; C. pepo seed, persimmon, pecan, Scirpus sp. seeds, stalk of Phragmites (reed grass).
3/4 mi. W. Monks Mound	Middle Mississippian	ca. AD 1000-1200 est.	Madison County	J. W. Bower, Jr.	12.0	6.3	2	—	—	100	—	—	
Mitchell, 20B-2-5	Middle Mississippian	ca. AD 1000-1200	Madison County	J. W. Porter	10.0	5.5	1	—	100	—	—	—	
Mitchell, 20-B-2-3	Middle Mississippian	ca. AD 1000-1200	Madison County	J. W. Porter	12.0	6.4	1	—	—	100	—	—	Many distorted corn grains, 8, 10, 12-rowed, about 2/3 medium, 1/3 small; Large quantity hickory nut shells, persimmon, wild cherry.
Wilson Mound (11SCOJ)	Middle Mississippian		St. Clair County	P. Holder	11.7	6.4	7	—	43	43	—	14	Hickory nut, pawpaw, pecan, grass seed, Equisetum, cattail.
McCain (20-B4-26)	Middle Mississippian (?)		Caseyville, St. Clair County	J. W. Porter									12 8-rowed corn grains, distorted, medium sized.

ILLINOIS

Site Name & No.	Culture/Tribe	Date	Location	Sample furnished by	Mean Row No.	Median Cupule Width (mm)	No. Cobs	Row Numbers % of Total Cobs						Other Plant Remains/Comments
								8	10	12	14	16+		
Texas No. 1 (21 B-3-6)	Middle Mississippian	C14 AD 1030-90±	Clinton County	R. Morrell	10.0	6.2	112	29	46	21	4	—	About 150 small to medium corn grains, a few probably 10-rowed, most could not be measured; hickory nut.
Schild	Mixed Middle Mississippian & Late Woodland	ca. AD 1050-1150	Green County	G. Perino									
Near Monks Mound	Stirling phase, Middle Mississippian	ca. AD 1050-1150	Madison County	G. Perino & J. Tarr	12.3	6.0	7	29	—	43	14	14	Corn grains; hickory nut, persimmon, pecan.
Ramey Farm, East Monks Mound	Stirling phase, Middle Mississippian	ca. AD 1050-1150	Madison County	G. Perino									Wild plum, persimmon, pawpaw.
Kellar (11MS99)	Mississippian	ca. AD 1050+		S. Denny	11.5	6.5	8	—	50	25	25	—	Hickory nut, persimmon.
Fingerhut, 1 1/2 mi. SW Monks Mound	Mississippian		St. Clair County	C. Bareis									One possible 8-rowed corn grain, medium-sized.
McDonough Lake, Bluff Site	Late Woodland & Middle Mississippian	ca. AD 1085	Madison County	G. Perino	11.2	5.9	5	—	40	60	—	—	
Kunneman Mound (11MS MD11)	Stirling phase, Middle Mississippian	est. AD 1000-1100		P. Holder	12.3	6.5	45	—	20	56	15	9	Also 8, 10, 12-rowed corn grains; hickory nut, persimmon, pawpaw, juniper, wild bean (S. helvola), grass and reed fragments.
C4-12	Mississippian		Wayne County	M. J. McNearney	8.7	7.4	3	67	33	—	—	—	
Theo. Arrel Farm	Mississippian		Jefferson County	J. Elliston									Hickory nut, black walnut
Mansker (24-A2 8)	Middle Mississippian (from wall trench house)	ca. AD 1010	Randolph County	J. W. Porter	11.2	6.8	36	17	22	44	17	—	
15B, 800 ft. W. Monks Mound	Late Woodland & Middle Mississippian	C14 AD 875-1545	Madison County	W. Wittry & R. Hall	10.7	6.9	64	13	48	31	6	2	Corn grains, persimmon, grass seeds
15A, 3000 ft. W. Monks Mound	Late Woodland & Middle Mississippian	C14 AD 825-1375	Madison County	W. Wittry & R. Hall	10.8	6.8	69	17	29	50	4	—	
J. Ramey Mound from depth of 8 ft.	Middle Mississippian		Madison County	E. Herold	16.0	6.0	1	—	—	—	—	1	More like "Mexican Dent" than "Tropical Flint". Not "NE Flint."

ILLINOIS

Site Name & No.	Culture/Tribe	Date	Location	Sample furnished by	Mean Row No.	Median Cupule Width (mm)	No. Cobs	8	10	12	14	16+	Other Plant Remains/Comments
Marty Coolidge (21 CL-18)	Mississippian	ca. AD 1000	St. Clair County	C. Kuttruff	10.9	6.3	28	21	25	43	11	—	Hickory and black walnut, persimmon
Cahokia (S-34-2), pit under Mound 51	Fairmount phase, Mississippian	ca. AD 1000+		C. Bareis & Wm. Chmurny	11.6	6.2	99	10	25	44	15	6	Large no. uncarbonized $C.$ $pepo$ seeds, $C.$ $pepo$ rind and peduncle var. $ovifera$ rind $Lagenaria$, chestnut burr, cane and juniper bark
Cahokia, Mound 72 (F-218)	Fairmount phase, Mississippian	C14 AD 999±60	St. Clair County	J. Anderson									4 corn cobs 12-14-rowed, C.W. 5.7-6.3 mm. Also 2 corn grains 6.5 mm wide and part of small shank (6.8 mm) probably from an ear not fully developed
Centerville (11S332)	Fairmount phase, Mississippian	ca. AD 950-1050		N. Lopinot & T. Norris									1 cupule from 12 or 14-rowed ear, cupule width 6.3 mm & 2 corn grains from 10 to 12-rowed ears; hickory nut, black walnut, wild grape, $Polygonum$ sp.
Horseshoe Lake	Mississippian	ca. AD 900-1100	Madison County	G. Perino	12.0	5.7	1	—	—	100	—	—	
Horseshoe Lake (E-73)	Fairmount phase, Mississippian		Madison County	M. L. Gregg	13.3	5.6	3	—	—	34	66	—	Also numerous corn grains, mostly small or medium sized; hickory nut, pecan, $Chenopodium$ or $Amaranthus$ sp.
Merrell Tract (11MS2-3)	Fairmount phase, Mississippian	ca. AD 900-1050		M. Woodworth	13.0	5.4	15	—	13	47	40	—	1 12-rowed cob is possibly popcorn. $C.$ $pepo$, persimmon, pecan, wild cherry
Kane (11MS194)	Late Woodland	AD 900-1050 (P. Munson)	Madison County	R. Hall	12.2	6	80	6	20	46	13	15	Corn grains, gourd, hickory nut, $Chenopodium$ sp., wild grape, grass seed, $Viburnum$, mallow
Powell Trent Tract (11MS2-2)	Late Woodland & Middle Mississippian	AD 620-1055 (C14)	Madison County	D. Lathrap	10.8	6.6	15	6	47	47	—	—	Slightly crescent-shaped, corn grains; hickory nut, persimmon, acorn, pecan, juniper
Grove Borrow Pit (11MS2-2)	Late Woodland & Middle Mississippian	AD 620-1055 (C14)	Madison County	C. Bareis	11.3	7.4	18	—	44	45	11	—	Corn grains, not crescent-shaped; black walnut, acorn, wild cherry, honey locust
Master F & S Co. (11S34-5)	Fairmount phase, Mississippian	C14 AD 850 ±210	St. Clair County	C. Bareis									8, 10, 12-rowed corn grains, distorted, mostly medium. A few small-sized; persimmon, wild cherry
Divers (11MO28)	Late Woodland & Mississippian	ca. AD 800-1000	Monroe County	G. Freimuth	11.7	6.5	36	3	22	53	22	—	Also corn grains. Sunflower, hickory nut, acorn, wild plum
Yokum	Late Woodland			B. Sudbury									Gourd, acorn
Stillwell	Late Woodland	ca. AD 600-800	Pike County	S. Struever									Hickory nut, pecan

ILLINOIS

Site Name & No.	Culture/Tribe	Date	Location	Sample furnished by	Mean Row No.	Median Cupule Width (mm)	No. Cobs	Row Numbers % of Total Cobs					Other Plant Remains/Comments
								8	10	12	14	16+	
11CW4	Middle and Late Woodland		Crawford County	Denzil Stephens									Hickory nut, Chenopodium or Amaranthus sp., Crataegus sp., Desmodium sp., sedge, (Carex sp.), rush, (Scirpus sp.)
Ansell (11CA17)	Hopewell?, possibly Late Woodland or Middle Mississippian		Calhoun County	J. C. McGregor									106 corn grains, mostly 12-rowed. A few 10-rowed or 14-rowed and 1 16-rowed
Apple Creek	Early Late Woodland, Whitehall phase	ca. AD 350-650	Greene County	S. Struever	12.0	5.1	1		—	100	—	—	Cob not carbonized; gourd, C. pepo
Kampsville Mound	Late Hopewell	ca. AD 400	Calhoun County	S. Struever									Hazelnut, honey locust seed
Peisker	Pike-Hopewell	AD 100-400	Calhoun County	S. Struever	12.5	5.6	8	—	—	75	25	—	Pop corn grains, probably 14-rowed
Snyders (11CAv8)	Hopewell & Late Woodland		Calhoun County	W. Fecht & S. Struever									Hickory nut, acorn, Chenopodium, Viburnum (?), unidentified grass seed
Jasper Newman (KS4) (F-13, F-18)	Middle Woodland	C14 80 BC ± 140 (M-1790)	Moultrie County	W.M. Gardner	10.0	7.0	9	45	22	22	11	—	Hickory nut
Macoupin Creek	Hopewell		Jersey County	S. Struever									1 12-rowed ear
Fry No. 1 (11JN96; 22C3-38)	Crab Orchard, Middle Woodland	ca. 100 BC - AD 100		E. Hall									Hickory nut, hazelnut
Consul (24B2-26)	Culture?	Date?	Jackson County	M. J. McNearney	9.3	7.8	9	45	44	11		—	

INDIANA

Site Name & No.	Culture/Tribe	Date	Location	Sample furnished by	Mean Row No.	Median Cupule Width (mm)	No. Cobs	Row Numbers % of Total Cobs					Other Plant Remains/Comments
								8	10	12	14	16+	
Rader	Upper Mississippian, Fisher & Oneota	ca. AD 1400-1600	Lake County	C. H. Faulkner									Tubers of white waterlily (Nymphaea tuberosa).
Greismer	Upper Mississippian, Fisher & Oneota	ca. AD 1400-1600	Lake County	C. H. Faulkner									Tubers of white waterlily (Nymphaea tuberosa), butternut, Prunus sp.
Angel Mounds	Middle Mississippian	ca. AD 1300-1500	Warrick County	J. H. Kellar	9.6	7.2	56	44	36	18	2	2	Also many corn grains, mostly medium-sized 8-10-rowed, usually wider than long. 12-rowed often longer than wide. Hickory nut, including some kingnut, wild plum, probably P. hortulana, black walnut, persimmon, pawpaw, pecan, grass, cane.
Farrand (12Vi64)	Mississippian	C14 AD 1085-1140		R. Pace	9.6	7.5	29	48	31	14	7	—	Also corn grains, some crescent-shaped. Common beans, median length 10.2 mm, median width 6.6 mm; pawpaw, hazelnut, possible wild cherry, ground nut (Apios).
Walsh-Dunlap (12Vi90)	Late Woodland			R. Pace	9.1	6.8	7	57	29	14	—	—	Hickory nut
Yankeetown	Late Woodland	ca. AD 800-1000	Warrick County	G. A. Black & K. Vickery									Distorted corn grains from 8, 10, and 12-rowed ears; hickory nut, black walnut, acorn, hazelnut
Daugherty-Monroe (12Su13)	Late Allison-LaMotte	UGA 1056 AD 355±70		R. Pace									19 C. pepo seeds, 8.5-11.9 mm long, 4.9-7.1 mm wide; acorn meats, possible locust (Robinia sp.)

IOWA

Site Name & No.	Culture/Tribe	Date	Location	Sample furnished by	Mean Row No.	Median Cupule Width (mm)	No. Cobs	Row Numbers % of Total Cobs					Other Plant Remains/Comments
								8	10	12	14	16+	
Malone II Shelter (13AM50)	Oneota, Upper Miss.	ca. AD 1600-1700		D. R. Henning	8.8	7.0	5	68	21	11	—	—	Also 14 corn grains. C. pepo, Prunus sp., hackberry, juniper.
Kimball (13PM4)	Mill Creek, Middle Mississippian	ca. AD 1100-1400		D. R. Henning	10.5	5.3	16	25	31	38	6	—	Some popcorn, about 17 corn grains, over half small 4.0 to 5.5 mm wide, the rest medium 6.5 to 7.8 mm wide.
Wittrock (13OB4)	Waterman Focus, Mill Creek, Middle Mississippian	ca. AD 1000-1500		D. R. Henning	12.0	5.0	2	50	—	—	—	50	The 16-rowed cob is popcorn; also one 8-rowed, large corn grain and one 10-rowed medium; sunflower, wild plum (P. americana).
13ML119	Nebraska phase	ca. AD 1100-1300		P. O'Brien	10.9	7.4	7	—	57	43	—	—	There were also distorted corn grains 7.0 to 9.5mm wide. The small cobs are probably popcorn. The large cobs have moderately open cupules and are moderately hardened.
13ML121	Nebraska phase	ca. AD 1100-1300		P. O'Brien	10.0	5.7	1	—	100	—	—	—	Also several dozen fragmentary corn grains; approx. 100 common beans, hickory nut, black walnut.
13CK15	Mill Creek			D. J. Stains									1 seed C. pepo similar to that of many small pumpkins, not cultivar "Mandan."
Phipps (13CK21)	Little Sioux Focus, Mill Creek, Middle Mississippian	ca. AD 900-1400		R. J. Ruppe & D. R. Henning	11.8	5.5	24	33	8	38	21	—	Some popcorn; C. pepo like "Red Lodge," wild plum, black walnut, hackberry, wild cherry.
13ML203	Nebraska phase	ca. AD 1100-1300		A. Anderson	10.0	8.4	2	—	100	—	—	—	
Beals (13CK62)	Corn in upper level-Great Oasis	ca. AD 1100-1200											No corn in lower levels; corn grains 8, 10, 12-rowed, most wider than long, width range, medium to large, 5.9 to 8.7 mm wide, no cobs; Lower levels - wild plum (P. americana), black walnut.
Meehan-Schell (13BN110)	Great Oasis	ca. AD 1000		B. Mead & D. A. Gradwohl	12.0	5.2	6	—	33	33	34	—	C. pepo, sunflower and Chenopodium sp. also said to be present.
Broken Kettle West (13PM25)	Great Oasis	C14 AD 880-1070 ±55		D. R. Henning & C. M. Johnson	12.9	4.7	27	—	15	22	45	7	Also corn grains from 8, 10, 12, & 14-rowed ears. Most from 10 & 12-rowed medium & small medium ears. Popcorn, C. pepo, sunflower, grape, hazelnut, Iva, Polygonum sp., hackberry, Scirpus sp., wild plum.

IOWA

Site Name & No.	Culture/Tribe	Date	Location	Sample furnished by	Mean Row No.	Median Cupule Width (mm)	No. Cobs	Row Numbers % of Total Cobs					Other Plant Remains/Comments
								8	10	12	14	16+	
Hadfield's Cave (13JN3)	Late Middle Woodland	C14 AD 800±55 (WISC-597) AD 295±65 (WISC-589)		D. Benn									Small to medium corn grain fragments; sunflower seeds reported by D. Benn.
Mound Group (13AM10)	Middle Woodland	C14 AD 160 ±300	Rathbun Res.	L. A. Brown thru E. W. Dodd									Inner bark, probably birch or alder
Woodpecker Cave (13JH202)	Middle Woodland		Coralville Res.	W. W. Caldwell thru E. W. Dodd						50	50	—	Hackberry, roots probably willow, stem of Composite similar to goldenrod
Near Glenwood, Iowa	Culture?	Date?		Brown & Anderson, Ames Museum	13.0		4	—	—				

KANSAS

Site Name & No.	Culture/Tribe	Date	Location	Sample furnished by	Mean Row No.	Median Cupule Width (mm)	No. Cobs	Row Numbers % of Total Cobs					Other Plant Remains/Comments
								8	10	12	14	16+	
Kansas Monument (14PR1)	Probably Republican Band, Pawnee	AD 1820-30		T. A. Witty	10.0	8.4	1	—	100	—	—	—	Also 65 corn grains, crescent-shaped, mostly from 8-rowed ears, some from 10-rowed & 2 12-rowed. 8-13 mm wide; approx. 100 common beans (P. vulgaris), 20 peas (P. savitum), wild plum (P. americana and P. hortulana).
Fanning (14DP1)	Kansa	ca. AD 1700		Wm. Bass & L. Hixon	9.0	11.0	2	50	50	—	—	—	ground nut, Apios americana
					9.0	grains	80	65	26	5	4	—	
El Cuartelejo (14SC1)	Plains, Apache, &/or Pueblo Refugees	ca. AD 1700		W. Wedel & T. A. Witty	10.6	8.0	14	29	43	7	14	7	Wild plum
Malone (14RC5)	Little R. Focus, Great Bend Aspect Wichita	ca. AD 1500-1600		W. Wedel	10.0	5.8	8	12	75	12	—	—	Also variable corn grains; some 7-10.5 mm wide are crescent-shaped and one 9.0 mm wide is wedge-shaped. Common beans.
Tobias (14RC8)	Little R. Focus, Great Bend Aspect Wichita	ca. AD 1500-1600		W. Wedel & J. M. Shippee	10.7	6.6	46	24	33	30	13	—	Most grains are not crescents, but are somewhat triangular, longer than wide, medium size. Many are thin. Wild plum.
Anthony (14HP1)	Big Bend Aspect (Caddoan influence)	ca. AD 900-1650		J. V. Chism	10.0	8.2	2	50	—	50	—	—	Also about 500 distorted grains, most from 8-rowed ears, but some from 10 and 12-rowed.
Squaw Creek (14DP25)	Nebraska Aspect	ca. AD 1500		J. M. Shippee	8.0	7.7	1	100	—	—	—	—	Common beans, some quite small.
Pratt (14PT1)	Great Bend Aspect	ca. AD 1400-1500		A. E. Johnson	10.0	5.2	2	50	—	50	—	—	Also corn grains from 8-rowed ears wider than long and others, mostly from 12-rowed, longer than wide.
Witt (14GE600)	Smoky Hill Aspect, Central Plains phase	ca. AD 1100-1450		P. J. O'Brien	11.6	7.1	5	—	40	40	20	—	Cupule on 14-rowed very open. Others open, but not as much. Black walnut.
Griffing (14RY401)	Smoky Hill Aspect, Central Plains phase	ca. AD 1100-1450		P. J. O'Brien	10.0	10.2	1	—	100	—	—	—	Cupules moderately open.
14PO4	Culture?	ca. AD 1199 ± 200		?									One possible corn grain; black walnut.
14CY102	Upper Republican	ca. AD 1200-1300		P. Holder	12.0	5.5	1	—	—	100	—	—	Also 8, 10, 12, 14-rowed grains, 3 large, 2 medium, and 5 small.
14ML11	Solomon R. variant, Upper Republican	ca. AD 1200-1300		P. Holder									Two or three large corn grains. Too broken to measure, one large and one medium.

KANSAS

Site Name & No.	Culture/Tribe	Date	Location	Sample furnished by	Mean Row No.	Median Cupule Width (mm)	No. Cobs	Row Numbers % of Total Cobs					Other Plant Remains/Comments
								8	10	12	14	16+	
Thull (14ML15)	Solomon R. variant, Upper Republican	ca. AD 1200-1300		P. Holder & M. D. Partsch	11.0	6.4	2	—	50	50	—	—	About 60 distorted corn grains, about 13 large, most medium and a few small. Sunflower, prairie turnip (Psoralea esculenta).
14ML16	Solomon R. variant, Upper Republican	C14 AD 1190 ± 90		P. Holder									Distorted corn grains and fragments. Mostly medium, a few small. C. pepo.
Long (14ML371)	Upper Republican	ca. AD 1200-1300		P. Holder									Distorted corn grains, mostly 8-rowed. Iva (?), hackberry.
Sumter (14OB27)	Upper Republican	ca. AD 1200-1300		P. Holder									Approx. 150 corn grains, a few large but mostly medium with a few small. Some wider than long, others longer than wide. 8-rowed and 10-rowed present.
Trowbridge (14WY1)	Kansas City Hopewell			N. R. Manion & E. Johnson	10.0	6.5	1	—	100	—	—	—	"Moderately thickened cob, less than northern flints, but not as open, early type cupule." H. C. C. Also several distorted grains probably from 8 & 10-rowed ears and, at least 1 from a 12-rowed, "flattened and longer than wide." C. pepo. persimmon, Iva.
Miller (14GE21)	Culture?	Date?		P. Holder & J. E. Sperry	12.0	6.3	1	—	—	100	—	—	
Rush Creek (14GE127)	Culture?	Date?		P. Holder & J. E. Sperry									Corn grains, one from 8-rowed ear, one from 10-rowed and three from 12 or 14-rowed. C. pepo, common bean.
14MH42	Culture?	Date?	Tuttle Creek Reser.	Carlyle Smith	8.0	7.9	1	100	—	—	—	—	

KENTUCKY

Site Name & No.	Culture/Tribe	Date	Location	Sample furnished by	Mean Row No.	Median Cupule Width (mm)	No. Cobs	Row Numbers % of Total Cobs						Other Plant Remains/Comments
								8	10	12	14	16+		
Fox Farm	Fox Farm Component, Madisonville, Ft. Ancient Aspect, Upper Miss. phase	ca. AD 1400-1600	Mason County	some material seen thru Geo. Carter									Both corn and common beans were present. See J. B. Griffin Fort Ancient Aspect, U. of Michigan Press, 1943.
Hardin Village (15GP22)	Ft. Ancient Aspect, Upper Miss.	after AD 1300		D. Schwartz	9.0	9.8	8	50	50			—	
15LY18A	Middle Miss.	after AD 1300		D. Schwartz	10.2	7.8	12	42	17	22	8	—	
Williams	Middle Miss.	Date?	Christian County	some material seen thru Geo. Carter									Corn cobs in burned post mold building. See William Site in Christian Co. Ky. by Webb and Funkhouser, U. of KY, Vol. 1, No. 1, July 1929.
Wickliff	Middle Miss.	est. AD 1200-1500	Ballard County	Brown & Anderson, Northern Flint Corns									"Corn with both 'Basketmaker and Eastern Complex affiliations'"
15CH2	Middle Miss.	after AD 1300		D. Schwartz	10.8	5.7	22	9	45	41	5	—	
15M14	Middle Miss.	after AD 1200		D. Schwartz	12.5	6.5	20	5	25	30	20	20	
15BT20	Middle Miss., perhaps peripheral	after AD 1200		D. Schwartz	11.8	6.5	8	—	37	38	25	—	
15CP3	Ft. Ancient Upper Miss.	after AD 1100		D. Schwartz	8.4	11.6	5	80	20	—	—	—	Corn may be from another site per Schwartz.
Steven Dehart Shelter (15PO1)	Cultural mtl. of several periods, some Adena			some material seen thru G. Carter & D. Schwartz	8.0	9.2	2	100	—	—	—	—	Not carbonized. Lagenaria.
Newt Kash Hollow	Cultural mtl. of several periods, some Adena		Menifee County	some material seen thru Geo. Carter	16.0		1	—	—	—	—	100	Not carbonized. Rind C. pepo, Lagenaria seeds and rind.
Salts Cave	Early Woodland	C14 BC 1190-280; C. pepo BC 400± 140 - BC 290± 200; Lagenaria BC 620± 140	Hart or Edmonson Co.	R. Hall & P. J. Watson									No Corn. Not Carbonized. C. pepo, some similar to variety ovifera; thick Lagenaria rind.
Floyd Day Shelter	Culture?	Date?		some material seen thru Geo. Carter									Not carbonized; 2 pieces thick C. pepo and Lagenaria rind.

KENTUCKY

Site Name & No.	Culture/Tribe	Date	Location	Sample furnished by	Mean Row No.	Median Cupule Width (mm)	No. Cobs	Row Numbers % of Total Cobs					Other Plant Remains/Comments
								8	10	12	14	16+	
15LE5	Culture?	Date?		some material seen thru Geo. Carter									Not carbonized: thick rinds from large Lagenaria.
Rogers Village Site	Culture?	Date?	Boone County	E. C. Crawford, Behringer Museum, Covington, KY									Not carbonized: acorn, seeds of Polygonum pennsylvanicum
Rogers Mound, Burial 27, Rock Tomb 4	Culture?	Date?	Boone County	E. C. Crawford									Wild cherry, wild grape, Viburnum sp., ragweed.

LOUISIANA

Site Name & No.	Culture/Tribe	Date	Location	Sample furnished by	Mean Row No.	Median Cupule Width (mm)	No. Cobs	Row Numbers % of Total Cobs					Other Plant Remains/Comments
								8	10	12	14	16+	
Jordon Site (16MO1)	Shell-tempered pottery	AD 1400+		R. W. Neuman	8.9	6.7	7	57	43	—	—	—	
Morton Shell Mound (16IB3)	Tchefuncte	BC 250 - AD 150		R. W. Neuman									Egg gourd (*C. pepo* var. *ovifera*), bottle gourd (*L. siceraria*). Also hickory nut, grape, acorn, persimmon, wild plum, *Polygonum* sp., *Viburnum* sp., *Smilax* sp., tupelo.

MICHIGAN

Site Name & No.	Culture/Tribe	Date	Location	Sample furnished by	Mean Row No.	Median Cupule Width (mm)	No. Cobs	Row Numbers % of Total Cobs					Other Plant Remains/Comments
								8	10	12	14	16+	
Ft. Michilimackinac	French inhabitant, Late British occupation	AD 1770 -1780	Emmet County	J. A. Brown	8.1	7.4	35	94	6	—	—	—	Also 8 corn grains, mostly swollen and distorted, medium sized, width greater than length; *C. pepo* seeds and peduncle, 2 seeds turnip, cabbage, or kale.
Dumaw Creek Site	Late Woodland	C14 AD 1680 ±75	Oceana County	G. Quimby									Squash (*C. pepo*) seeds 12.0 x 7.5 mm, probably summer squash or a small pumpkin.
Butler	Probably related to Owasco	ca. AD 1500-1650 est.	Wayne County	A. R. Woolworth									Corn grains 89 8-rowed, 2 10-rowed, 12 8 or 10-rowed, but probably most 8-rowed, crescent-shaped. No small grains like popcorn.

MINNESOTA

Site Name & No.	Culture/Tribe	Date	Location	Sample furnished by	Mean Row No.	Median Cupule Width (mm)	No. Cobs	Row Numbers % of Total Cobs					Other Plant Remains/Comments
								8	10	12	14	16+	
Ft. Renville	Historic American Fur Co. post	AD 1825-1846		D. Nystuen	8.2	9.4	11	91	9		—	—	
Burial mound at Mill Lacs Lake (21ML16)	Proto-historic Eastern Dakota			E. Johnson									Approx. 65 whole, uncarbonized seeds C. pepo and frags. of more. Range 5.2 x 10.3 to 7.7 x 15.5, median 7.0 x 13.3 mm. Seeds Prunus sp., probably P. pennsylvanicum amd grape. Vitis sp.
Bryan	Mississippian with Oneota sherds present	C14 AD 1120		G. Gibbon	8.0	9.1	1	100	—	—	—	—	
Nelson (21BE24)	Probably Late Woodland	ca. AD 1000		M. Scullin	11.2	5.9	5	20	—	80	—	—	Also corn grains from 8, 10, and 12-rowed ears, most longer than wide.

Archaeological Sites East of the Rockies

MISSISSIPPI

Site Name & No.	Culture/Tribe	Date	Location	Sample furnished by	Mean Row No.	Median Cupule Width (mm)	No. Cobs	Row Numbers % of Total Cobs					Other Plant Remains/Comments
								8	10	12	14	16+	
Grand Natchez Village	Proto-Historic & historic Natchez	FIND 372 = ca. AD1582±; FIND 327=ca. AD 1682±	Adams County	R. S. Neitzel	F 372 9.4	6.1	21	33	62	5	—	—	C. moschata peduncle (FIND 148, Historic period AD 1700-1729)
					F 327 12.0		4	—	—	100	—	—	
County Hospital	Proto-historic Chickasaw or Chocchuma	pre-AD 1700	Oktibbeha County	R. A. Marshall	11.5	7.6	8	—	50	25	25	—	Hickory nut, persimmon, acorn, pecan, Ampelopsis
Powell Bayou (22SU516)	Mississippian, Middle to Late	ca. AD 1200-1500		J. M. Connaway & R. A. Marshall	10.5	5.5	31	16	52	26	6	—	2 common beans, persimmon, wild sunflower (?), honey locust.
Wilford (22CO516)	Mississippian	ca. AD 1200-1500		J. M. Connaway	11.2	4.5	5	20	20	40	20	—	3 cupules, 8 and 10 rowed, cupule width 7.3 to 9.0 mm.
Buford	Mississippian	ca. AD 1200-1500+	Tallahatchee County (P., F., Griffin #1701)	R. A. Marshall									
Lyons Bluff (22OK1)	Mississippian	ca. AD 1200-1500		R. A. Marshall	10.8	6.1	78	10	46	39	4	1	Common beans
Bobo (22CO535)	Mississippian	C14 AD 1275±100		J. Connaway									Approx. 75 corn grains. Most from 12-rowed ear or ears. Most appear longer than wide.
Hays (22CO612)	Mississippian	ca. AD 1000-1200		J. Connaway	11.6	6.4	14	7	29	50	7	7	Also corn grains. Hickory nut, persimmon
Humber (22CO601)	Mississippian			D. L. Fichtner & J. Connaway	11.3	5.9	9	56	33	11	—	—	Also corn grains. Common beans, persimmon, pecan, possible sunflower
Lafferty West (22CO636)	Mississippian			J. Connaway	10.0	5.0	—	—	100	—	—	—	Also corn grains, apparently from same ear, 5 to 7 mm wide.
Flower No. 3 (22TU518)	Mississippian			J. Connaway	10.6	5.9	22	14	41	45	—	—	Corn grains 5 to 8 mm wide. Half common bean, hickory, persimmon, pecan, wild grape, Strophostyles sp., Scirpus or Carex sp.
Craig (22CO566)	Mississippian, Early to Middle			J. M. Connaway	11.3	5.3	15	13	27	40	20	—	Hickory nut
Bonds (22TU530)	Late Baytown & Early Mississippian	ca. AD 800-900		J. M. Connaway	12.0	6.4	7	—	29	43	28	—	Prunus sp., hickory nut, persimmon, acorn, pecan, butternut
Noe (22CO587)	Early to Middle Baytown	ca. AD 500-700		J. M. Connaway									Black walnut, persimmon, pecan, Ampelopsis, Strophostyles umbellata, chokecherry (Prunus sp.)

MISSISSIPPI

Site Name & No.	Culture/Tribe	Date	Location	Sample furnished by	Mean Row No.	Median Cupule Width (mm)	No. Cobs	Row Numbers % of Total Cobs					Other Plant Remains/Comments
								8	10	12	14	16+	
Acree (22BO551)	Baytown-Marksville			J. Connaway									Hickory nut, persimmon, acorn, pecan
Boyd (22TU532)	Tchula, Late Marksville, Early Baytown			J. Connaway									Hickory, black walnut, persimmon, acorn, pawpaw, pecan, Chenopodium sp., wild grape, marshelder (Iva).
Teoc Creek (22CR504)	Poverty Point	BC 1000-800		J. M. Connaway									Black walnut or butternut
Denton (22QU522)	Poverty Point Relationships			J. M. Connaway									Hickory nut, black walnut, persimmon, acorn, butternut
Longstreet (22QU523)	Middle Archaic			J. M. Connaway									Hickory, black walnut, persimmon, acorn, pecan, possible Strophostyles sp.

MISSOURI

Site Name & No.	Culture/Tribe	Date	Location	Sample Furnished by	Mean Row No.	Median Cupule Width (mm)	No. Cobs	8	Row Numbers % of Total Cobs 10	12	14	16+	Other Plant Remains/Comments
Schroer Rock Shelter	Probably modern		Montgomery County	C. K. Sheets Jr.	15.6	6.5	9	—	—	22	22	56	
Carlton Woodward (23RIHI)	Historic Pioneer Squatter	ca. AD 1815-1845		J. & C. Price	12.4	5.0	5	—	40	20	20	20	2 elongated corn grains, 7.7 & 11 mm long, which are southern dent corn. 10-rowed cobs are eastern 8-rowed and 12, 14, 16-rowed cobs are popcorn. Small common beans, watermelon, peach, hickory, wild plum, black walnut, persimmon, hackberry, hazelnut, wild grape, hawthorn.
Carrington (23VE1)	Historic Great Osage	ca. AD 1785-1820		C. H. Chapman	8.9	8.7 On 30 cobs.	69	67	23	10	—	—	On 30 cobs.
Coal Pit (23VE4)	Historic Little Osage	AD 1790-1815		C. H. Chapman	8.9	9.0	328	63	30	6	1	—	On 246 cobs. _C. pepo_, common beans, peach, watermelon, cultivated cherry (_P. avium_), hickory nut, acorn, persimmon, hackberry, pawpaw, lotus.
Brown (23VE3E)	Historic Osage	ca. AD 1700-1800		C. H. Chapman	9.6	9.0	15	47	33	13	7	—	Hickory nut, black walnut or butternut _Euonymous_ sp.?
Utz (23SA2)	Upper Mississippian, Historic Missouri	ca. AD 1650-1730		C. H. Chapman & R. T. Bray	9.4	8.5	66	47	38	13	2	—	Common beans, median length 11.6 mm, median width 6.3 mm, _C. pepo_, watermelon, hickory, wild plum, black walnut, persimmon, hackberry, pawpaw, pecan, hazelnut, wild sunflower, wild grape, wild crabapple, hawthorn, _Rubus_ sp.
King Hill (23BN1)	Late Oneota, Kansa?	ca. AD 1670-1720		D. R. Henning et.al.	9.8	8.2	742	42	32	19	5	2	_C. pepo_, common bean, _Lagenaria_, watermelon, peach, hickory, wild plum, black walnut, acorn, hackberry, pawpaw, hazelnut, wild grape, _Rubus_ sp.
Rock Shelter (23BY530)	Marginal Middle Miss. & top layers Neosho focus or early Osage & top layers Neosho focus or early Osage	ca. AD 1400-1800?		C. H. Chapman	CN19= 9.5 Other= 10.4	9.2 8.0	16 28	37 36	50 28	13 21	— 11	— 4	Kernels dented & not dented. Not carbonized. Common beans, hickory nut, wild plum, black walnut, acorn, pawpaw, hazelnut, sunflower, buffalo berry

MISSOURI

Site Name & No.	Culture/Tribe	Date	Location	Sample Furnished by	Mean Row No.	Median Cupule Width (mm)	No. Cobs	Row Numbers % of Total Cobs					Other Plant Remains/Comments
								8	10	12	14	16+	
Cloud Williams	Neosho or earlier		McDonald County	G. Perino from D. Dickson	12.0	5.0	1	—	—	—	—	—	persimmon
Hess (23Ml55)	Mississippian	ca. AD 1400-1600		R. B. Lewis	12.0	5.4	2	—	—	100	—	—	Also 10 and 12-rowed corn grains. Persimmon, acorn, pawpaw, pecan, Vitis sp., Viburnum sp., Ampelopsis sp.
Callahan-Thompson (23Ml71)	Mississippian	ca. AD 1400-1600		R. B. Lewis	10.8	6.0	55	16	33	44	7	—	Common beans, hickory nut, kingnut, black walnut, persimmon, acorn, pawpaw, pecan, wild cherry, Viburnum sp., Strophostyles sp., Ampelopsis sp., Passiflora sp., coffee tree.
Lilbourn (23NM38)	Mississippian & Baytown			R. Pangborn & J. W. Cottier	10.7	5.3	66	23	28	36	13	—	Popcorn, common beans, hickory nut, wild plum, black walnut, persimmon, acorn, hackberry, pawpaw, pecan, sunflower, wild grape, Iva, Viburnum sp., black gum.
House on Mound, North end Lilbourn (23NM49)				J. W. Cottier									Persimmon, acorn.
Byrd (23Ml53)	Mississippian & small amount Baytown			J. W. Cottier									2 corn grains. Common bean, hickory nut, black walnut, persimmon, Apios.
Guthrey (23SA131)	Upper Mississippian Oneota	ca. AD 1200±80-1450±60		D. R. Henning	9.4	7.5	7	56	14	29	—	—	Distorted kernels, some crescent. Wild plum, persimmon, hazelnut, basswood.
Crosno (23Ml1)	Late Middle Mississippian-Cairo Lowlands phase, New Madrid Focus	ca. AD 1200-1400 est.		S. Williams	10.1	6.8	18	28	39	33	—	—	
Boyce Mound (23JO40)	Plattin Focus, Mississippian	ca. AD 1200-1400 est.		R. M. Adams, AC, SC, StL.	cob=12 11.1	5.2 grains	1 cob 34	— 18	— 32	100 35	— 9	— 6	Scirpus sp.

Archaeological Sites East of the Rockies 121

MISSOURI

Site Name & No.	Culture/Tribe	Date	Location	Sample Furnished by	Mean Row No.	Median Cupule Width (mm)	No. Cobs	Row Numbers % of Total Cobs					Other Plant Remains/Comments
								8	10	12	14	16+	
Turner-Snodgrass (23BU21)	Mississippian	ca. AD 1300		J. Price	11.0	6.3	51	6	49	35	8	2	Also corn grains. Hickory nut, black walnut, persimmon, acorn, sunflower, Iva
Gypsy Joint (23RI101A)	Powers phase, Mississippian	ca. AD 1300		B. D. Smith									3 vials of 10 and 12-rowed corn grains. These are small, probably flint, flattened and probably from small ears. 3 measured, 1 from 10-rowed ear 6.7 mm, wide, 7.2 mm long, 1 from 12-rowed, 6.0 x 8.1 x 3.6 mm, and another from 12-rowed, 5.6 x 6.2 x 3.1 mm
23MI501	Mississippian	ca. AD 1200-1300		J. W. Cottier	12.0	5.4	2				100		Persimmon, Apios americana, goldenrod, gall.
23MI106	Mississippian	ca. AD 1200-1300		J. W. Cottier	12.3	6.5	7		14	57	29		Wild plum, hawthorn (Crataegus sp.).
Towosahgy (23MI2)	Mississippian	ca. AD 1200-1300		J. W. Cottier	10.8	6.2	75	17	41	28	3		Common bean, hickory nut, black walnut, persimmon, acorn, pecan, Iva, ragweed, Passiflora sp., virginia creeper, wild plum, Apios.
23MI59	Mississippian			J. Price & R. Williams									Persimmon seeds and fruit
Vandiver Mound (23PL6 Md. B)	Mississippian	C14 AD 1290±80		J. M. Shippee									8, 10, and some 12-rowed kernels. 8 & 10 rowed northern flint, 12-rowed probably small flint. may be old or may be more recent sample of pop ears.
McClarnon (23PL54)	Mississippian	C14 AD 1260±90 (Square wall trench house)		J. M. Shippee	9.0	8.0	4	50	50				
Near Mouth James River (23SN42)	Mississippian relations with Gibson Aspect	est. ca. AD 1100±400		W. R. Wood									12 - 14F cob or cobs. C. W. 7.0 mm. (H. C.) "may be mixture of early corn with N. flint."

MISSOURI

Site Name & No.	Culture/Tribe	Date	Location	Sample Furnished by	Mean Row No.	Median Cupule Width (mm)	No. Cobs	8	10	12	14	16+	Other Plant Remains/Comments
Friend and Foe (23 CL113)	Mississippian	C14 (M-9026) AD 1100±110		F. A. Calabrese	cob 12	6	1					—	8, 10, & a few 12-rowed corn grains and cupules. Northern flint, but not an extreme form. Most cupules are narrow, but some are open part. on higher row numbered cobs. *C. pepo*, hickory nut, black walnut, *Iva*.
Jep Long (23JV35)	Mississippian	est. ca. AD 1000 1200		Exc. R. M. Adams, Acad. of Science of St. Louis	9.1	grains	33	61	24	15	—	—	
Antire Creek (23SL62)	L. Woodland & Mississippian (Corn from Miss.)	est. ca. AD 1000 1200+		L. W. Blake	10.2	6	22	18	59	18	5	—	Based on 3 measurable cobs 8, 10, 14-rowed and 19 kernels.
Double Bridges (23NM154)	Baytown & Mississippian			J. W. Cottier									Hickory nut, *C. ovata* and *C. laciniosa*, acorn, persimmon.
Young (23PL4)	Steed-Kisker Miss.	ca. AD 1000-1200		P. J. O'Brien	10.2	7.1	9	22	45	33	—	—	Also large, medium, & small corn grains. Common beans, hickory nut, hazelnut, sunflower, (cult. & probably wild), *Iva*, wild crabapple(?).
Pridey (23PL14)	Steed-Kisker Miss.	ca. AD 1000-1200		P. J. O'Brien	8.7	7.2	3	67	33	—	—	—	Also fragments of corn grains. 33 common beans, black walnut, acorn.
Coons (23PL16)	Steed-Kisker Miss.	ca. AD 1000-1200		P. J. O'Brien	9.6	7.0	9	45	33	22	—	—	Also rather large corn grains. *C. pepo*, common beans, hickory nut, hazelnut, *Iva*.
White (23PL80)	Steed-Kisker Miss.	ca. AD 1000-1200		P. J. O'Brien	10.0	7.5	7	29	57	—	14	—	Also crescent-shaped corn grain 7.3 mm wide.
McColloch (23NM252)	Early Mississippian	AD 1000±400		R. A. Marshall									Sample 30 kernels: 17 8-rowed, 10 10-rowed, 3 12-rowed. Eight-rowed wider than long, 12-rowed longer than wide. Three quarts distorted kernels. Persimmon, *Iva*.

MISSOURI

Site Name & No.	Culture/Tribe	Date	Location	Sample Furnished by	Mean Row No.	Median Cupule Width (mm)	No. Cobs	\multicolumn{5}{c}{Row Numbers % of Total Cobs}	Other Plant Remains/Comments
								8	10	12	14	16+	
Reserach Cave (23CY6)		Organic matter below 12-24" probably to AD 1100		exc. J. M. Shippee, C. H. Chapman	16.0	5.5	2	—	—	—	—	100	Not carbonized. *C. pepo*, *Lagenaria*, pawpaw, honey locust, *Cirsium* leaf
Steed-Kisker (23PL13)	Mississippian, Steed-Kisker focus	C14 AD 1080±80		J. M. Shippee	9.0	6.8	2	50	50			—	
Meyer	Mississippian & some Late Woodland	est. ca. AD 900-1200	St. Louis County	W. O. Meyer									One pc. ear 8-10-rowed; one pc. ear 10-12 rowed; one pc. ear 12-rowed
River Bend East, House No. 1 (23SL79)	Late Woodland & Mississippian	est. ca. AD 800-1200		D. R. Henning	10-11	8.1	2		50?	50	—	—	Corn grains from 8, 10, and 12-rowed ears, not extreme crescent-shaped.
River Bend East, Pit (23SL79)	Late Woodland & Mississippian	est. ca. AD 800-1200		D. R. Henning				50?	50?				6 corn grains from 8-rowed ear. 2 from a 10-rowed. Most are crescent-shaped, rather small; but 1/2 to 1/3 longer than usual for northern flint.
23H208	M. Mississippi	ca. AD 1000 est.	Pomme de Terre Reservoir	R. T. Bray									13 8-rowed, 2 10-rowed kernels, medium 8.0-9.5 mm wide.
Madrigal Mound (23PO300)	Mississippian	est. ca. AD 1000		W. R. Wood	8.0	11.8	1	100	—	—	—	—	8, 10, 12 kernels, mostly medium, some small, a few large.
King's Curtain Mound (23PO307)	Mississippian	est. ca. AD 1000		W. R. Wood									Three quarts corn kernels, too distorted for measurments. Mostly 8, 10-rowed, a few 12-rowed. Hickory and hazelnuts.
23DA225	Mississippian			D. R. Henning									Distorted kernels, most 10-rowed, not crescent-shaped. Some medium 8-rowed, crescent-shaped. A few 12-rowed, medium-sized, not crescent-shaped
Warren Gresham (23PL48)	M. Mississippian	C14 AD 875±150 (rectangular house)		J. M. Shippee	12.0		1	—	—	100	—	—	Cob rotted. Kernels small, slightly large and flattened for pop.

124 *Archaeological Sites East of the Rockies*

MISSOURI

Site Name & No.	Culture/Tribe	Date	Location	Sample Furnished by	Mean Row No.	Median Cupule Width (mm)	No. Cobs	Row Numbers % of Total Cobs					Other Plant Remains/Comments
								8	10	12	14	16+	
Rock Shelter	Late Woodland		Warren County	W. Barnes									Corn cobs-not carbonized-probably modern. <u>C. pepo</u>, pumpkin type and Mandan type. Sunflower, hickory nut.
23FR124	Late Woodland			F. Schneider									Hickory nut, black walnut
23FR234	Late Woodland			F. Schneider									Hickory nut, black walnut, hazelnut
Rock Shelter (23HI134)	Late Woodland?			W. R. Wood	10.0	11.7	1		100	—	—	—	Not carbonized.
Popke Village (23MN302)	Early Late Woodland	ca. AD 800 or <u>earlier</u> (Weaver ware present)		D. R. Henning									20 Kernels. 2/3 medium 10-12-rowed, 1/3 small 12-14-rowed, thin and narrow. These suggest hard flint with unusually long grains. Majority medium-small flints.
Kersey (23PM42)	Corn believed to be from Baytown occup.	700AD±200		R. A. Marshall									From 12 to 14-rowed ears -- small flint or medium pop. Distorted kernels.
Hoecake (23MI8)	Early to Late Baytown	500-900 AD		R. Williams	11.3	5.2	3	—	33	67	—		Cupules only. Also some corn grains. Persimmon, acorn, pecan, wild grape, and unidentified tubers
Boney Spring Pit		C14 AD 30±50 (TX-1472)	Benton County	F. B. King									<u>Not carbonized</u>. Squash (<u>C. pepo</u>) like cultivar "Mandan."
Tom Baker	Late Early Woodland	est. ca. BC 100	Stoddard County	R. A. Marshall									Two 8, 10-rowed corn kernels. Hickory, black walnut, persimmon, pecan
23MN223	Early Woodland	C14, BC 570±150 & BC 660±200		W. Klippel									Hickory nut, black walnut, <u>Chenopodium</u>, wild grape, <u>Iva</u>, <u>Pyrus ioensis</u> (?).
Phillips Spring (23HI216)		Average C14 2330 BC ±50 (SMU-98-102)		F. B. King									<u>Not carbonized</u>. Squash (<u>C. pepo</u>) like cultivar "Mandan."
23MN255	Culture?	Date?		W. Klippel									One 8-rowed corn grain and fragments.

MISSOURI

Site Name & No.	Culture/Tribe	Date	Location	Sample Furnished by	Mean Row No.	Median Cupule Width (mm)	No. Cobs	Row Numbers % of Total Cobs					Other Plant Remains/Comments
								8	10	12	14	16+	
Rockshelter (23MD43)	Woodward Plain & Neosho Punctate ceramics, but cultural assoc. not always clear			J. Cobb									2 cobs of 14-rowed popcorn, cupule width 4.0 and 4.3 mm. 1 cob of 10-rowed ear, which could be modern or old. Gourd, watermelon, persimmon, hackberry, Viburnum, locust.
Dry cave	Culture?	Date?	Pulaski County	J. F. Keefe	10.0	8.4	1	—	100	—	—	—	Not carbonized.
Rock Shelter (23BY471)	Culture?	Date? Probably modern 1800+		C. H. Chapman	12.2	8.0	25	16	16	24	32	12	Not carbonized. Shoe-peg dent kernels. Hickory nut, wild plum, black walnut, persimmon, acorn, hackberry, hazelnut, Iva, Viburnum, ragweed, buffalo berry, pine, redbud, morning glory.
Rock Shelter (23BY476A)	Culture?	Date? Probably modern		C. H. Chapman	12.0	9.8	2	—	—	100	—	—	Not carbonized. Hickory nut, black walnut, acorn, hackberry, pecan (?).
Rock Shelter (23BY476B)	Culture?	Date? Probably modern		C. H. Chapman	11.0	11.1	2	50	—	—	50	—	Not carbonized. Wild plum, hickory nut, acorn, hackberry, pawpaw, Iva, buffalo berry, redbud.
Rock Shelter (23BY485)	Culture?	Date?		C. H. Chapman									Acorn, wild grape, ragweed
Rock Shelter (23BY522)	Culture?	Date?		C. H. Chapman	9.7		6	17	83	—	—	—	Not carbonized. Prunus sp., hackberry
Rock Shelter (23BY529)	Culture?	Date? (Disturbed area?)		C. H. Chapman									C. pepo, Lagenaria, wild plum, acorn, juniper
Rock Shelter (23BY531)	Culture?	Date?		C. H. Chapman	9.0	7.0	2	50	50	—	—	—	Not carbonized. Kernels 10-rowed. Sunflower.

NEBRASKA

Site Name & No.	Culture/Tribe	Date	Location	Sample Furnished by	Mean Row No.	Median Cupule Width (mm)	No. Cobs	Row Numbers % of Total Cobs					Other Plant Remains/Comments
								8	10	12	14	16+	
25ST1	Historic Omaha Oneota (Iowa)	AD 1820-25 ca. AD 1650-1700		J. L. Champe	8.9	9.2	57	58	40	2	—		Also corn grains, mostly from 8 and 10-rowed ears and about 50 small grains from 14-rowed, probably popcorn. C. pepo like "Mandan" and others, wild plum, ragweed.
Hill (25WT1)	Historic Pawnee	AD 1800		J. L. Champe									Charred corn husk cordage
Ponca Fort (25KX1)	Aksarben Ponca	AD 1300-1400 AD 1790-1800		J. L. Champe									Corn grains from 8 and 10-rowed ears. Wild plum, wild cherry, rush, cottonwood bark
25DK5	Historic Omaha	AD 1770-1800		J. L. Champe	9.0	10.1	16	50	50		—		—
25BD1	Aksarben Oneota (Iowa)	ca. AD 1400 AD 1650-1700		J. L. Champe									Charred grass.
25CX2	Lower Loup R. Complex, Early Pawnee	ca. AD 1650-1700		J. L. Champe									Black walnut, hazelnut.
Minaric (25K9)	Redbird Focus (Late)	ca. AD 1650-1700		J. L. Champe	8.0	7.5	1	100			—		wild plum, Lithospermum sp.
Redbird I (25HT3)	Redbird Focus	AD 1600-1700		J. L. Champe									Wild plum, hackberry
25CC30	Aksarben	ca. AD 1300-1400		J. L. Champe									Three corn grains from 8-rowed ear.
25CD1	Aksarben	ca. AD 1300-1400		J. L. Champe									"Charred, woven grass."
Bobier (25DK1)	Aksarben	ca. AD 1300-1400		J. L. Champe, D. J. Blakeslee									2 cobs from 10-rowed ears. Cupule widths 6.2 and 7.4 mm.
25DK7	Aksarben	ca. AD 1300-1400		J. L. Champe, D. J. Blakeslee									Corn grains, common beans.
25DX1	Aksarben	ca. AD 1300-1400		J. L. Champe, D. J. Blakeslee							—		Charred corn grains, mainly from 8-rowed ears, some from 10-rowed and a few from 12-rowed.
Walker Gilmore (25CC28)	Aksarben	ca. AD 1400		J. L. Champe	14.0	4.0	1				100		Also a large northern flint corn grain from 8-rowed ear. Wild plum.

NEBRASKA

Site Name & No.	Culture/Tribe	Date	Location	Sample Furnished by	Mean Row No.	Median Cupule Width (mm)	No. Cobs	Row Numbers % of Total Cobs					Other Plant Remains/Comments
								8	10	12	14	16+	
25TS11	Nebraska Aspect			R. D. Grant									14 distorted corn grains, eight from 8-rowed and one from 10-rowed ears. One broken common bean and unidentified legume.
Goodrich (25GY21)	Upper Republican probably Loup River phase	ca. AD 1200-1500		W. J. Hunt	10.5	6	4	50	—	25	25	—	The 14-rowed cob may be popcorn. Common bean, sunflower, Iva.
25FT35	Upper Republican	ca. AD 1200		R. W. Wood	11.6	6.2	5	—	40	40	20	—	Also two small distorted corn grains.
Walker Gilmore (25CC28)	Sterns Creek	ca. AD 900		J. L. Champe, P. Holder									Small seeded C. pepo, Lagenaria, hickory nut, black walnut, prairie turnip (Psoralea esculenta).

SELECTED SITES OF EASTERN NEW MEXICO

Site Name & No.	Culture/Tribe	Date	Location	Sample Furnished by	Mean Row No.	Median Cupule Width (mm)	No. Cob	Row Numbers % of Total Cobs					Other Plant Remains/Comments
								8	10	12	14	16+	
San Lorenzo Picuris Pueblo	Penasco	AD 1930 to present	Taos County	J. Dick	14.0		40	—	8	32	15	45	Not carbonized.
Picuris Pueblo	Trampas Phase	AD 1600-1696	Taos County	H. Dick	11.0	5.4	324	18	28	44	7	3	(70 cobs) C. maxima, peach, common bean
Gran Quivera Natl. Monument	Peripheral Anasazi	Mostly AD 1600-1672	Torrance & Socorro Counties	A. Hayes	11.1	5.2*	1309	12	32	44	10	2	*(Limited Sample). C. pepo, common bean, Lagenaria, cotton, Amaranthus.
Picuris Pueblo	Valido & San Lazaro phases	AD 1375-1600	Taos County	H. Dick	10.5		26	19	42	35	—	4	C. pepo, C. mixta
Las Madres (LA25)	Pueblo III-IV	ca. AD 1250-1350	Galisteo Basin	B. P. Dutton	11.6		135	7	27	49	13	4	
Pueblo Largo (LA183)	Pueblo III-IV	ca. AD 1250-1350	Galisteo Basin	B. P. Dutton	11.6		175	8	20	58	11	3	Lagenaria
Picuris Pueblo	Taos & Santa Fe phases	AD 1150-1300	Taos County	H. Dick	10.3		35	34	26	31	6	3	C. pepo
Six Sites	Largo-Gallina phase	ca. AD 1100-1300	Rio Arriba County	R. C. Green	11.6		171	7	26	45	19	3	C. mixta, C. pepo?
Hatchet (29OT-3)	Northern Jornada Branch Mogollon	ca. AD 1050	Otero County	E. B. McCluney	10.2	7.6	10	40	20	30	10	—	Mesquite pods

Archaeological Sites East of the Rockies   129

NEW YORK

Site Name & No.	Culture/Tribe	Date	Location	Sample Furnished by	Mean Row No.	Median Cupule Width (mm)	No. Cobs	Row Numbers % of Total Cobs					Other Plant Remains/Comments
								8	10	12	14	16+	
Reed Fort, NY	Iroquois			C. H. Hayes, III	8.0	7.9	1	100	—	—	—	—	Also large, crescent-shaped corn grains.
Cornish W. Bloomfield (HNE9-2)	Early Historic Iroquois	AD 1600	Ontario County	C. H. Hayes, III									Corn grains, predominately large, 8-rowed. Common beans, wild plum, hickory nut, acorn
Alhart	Prehistoric Iroquois	AD 1400-1600	near Churchville, Monroe County	C. H. Hayes, III	8.1	11.1	14	93	7	—	—	—	Also one bushel of corn grains, mostly 8-rowed.
Can. (29-3)	Prehistoric Iroquois, probably Seneca	AD 1300-1500	near Bristol	C. H. Hayes, III	8.3			85	15	—	—	—	(Corn grains 27), Common beans, wild plum, hickory nut. Vaccinium sp.
Sackett	Owasco	AD 1065-1215	Ontario County	Brown & Anderson Rochester Museum	8.2		22	91	9	—	—	—	Common beans
Silverheels	Owasco		Erie County	Brown & Anderson Peabody Museum, Harvard	9.0		2	50	50	—	—	—	

NORTH DAKOTA

Site Name & No.	Culture/Tribe	Date	Location	Sample Furnished by	Mean Row No.	Median Cupule Width (mm)	No. Cobs	Row Numbers % of Total Cobs					Other Plant Remains/Comments
								8	10	12	14	16+	
Ft. Berthold Village (32ML12)	Historic Mandan, Hidatsa, Arikara	AD 1845-74		P. A. Ewald	8.8	9.4	5	60	40	—	—	—	Large crescent-shaped corn grains. *C. pepo*, *C. maxima*. *Prunus americana*
Deapolis Site (32ME5)	Historic Mandan	Occupied until AD 1838-1840		W. R. Wood	8.5	8.8	4	75	25	—	—	—	Crescent-shaped, northern flint corn grains. *C. pepo*, wild plum.
Rock Village (32ME15)	Historic Hidatsa	Probably about AD 1835-37		D. D. Hartle									*C. pepo*, *C. moschata*
Sacajawea (32ME11)	Historic Hidatsa	early 19th century		J. J. Hoffman									*C. pepo* seeds, "quite uniform, like small and slightly immature Mandan"
Boley (32MO37)	Protohistoric Mandan, said to have been abandoned	AD 1764		W. R. Wood, D. Lehmer	8.4	9	17	82	18	—	—	—	*C. pepo*, "summer squash" and "Mandan" varieties
Motsiff	Protohistoric Mandan	ca. AD 1700-1750		P. A. Ewald	9.4	7.3	7	43	43	14	—	—	
Sperry (32B14)	Protohistoric Mandan	ca. AD 1700-1750		W. R. Wood	9.0	7.2	6	66	17	17	—	—	
Mahhaha (32OL22)	Protohistoric Hidatsa			W. R. Wood									From "medicine bundle", tobacco seeds (*N. rustica*) and plum stone (*P. americana*).
Double Ditch (32B18)	Heart River Focus, Protohistoric Mandan	ca. AD 1700-1725		M. Thurman	8.6	8.1	29	80	10	10	—	—	Cucurbit rind.
Shermer (32EM10)	Huff Focus, Terminal Horizon, Middle Mo. Tradition	ca. AD 1485-1543		J. E. Sperry	10.0	5.9	20	40	25	30	5	—	*C. pepo*, common bean
Paul Brave (32S14)	Extended Middle Mo. Tradition	est. AD 1350-1450		W. R. Wood	10.0	8.0	2	—	100	—	—	—	Northern flint corn grains. *C. pepo*, "Mandan" variety.

OHIO

Site Name & No.	Culture/Tribe	Date	Location	Sample Furnished by	Mean Row No.	Median Cupule Width (mm)	No. Cobs	Row Numbers % of Total Cobs 8	10	12	14	16+	Other Plant Remains/Comments
Garriel	Baum Focus, tr. Ancient Aspect	ca. AD 1500-1600	Athens County	J. L. Murphy	8.0	9.4	1	100	—	—	—	—	Common beans, wild plum, hazelnut
South Park	Whittlesey Focus	ca. AD 1500+	Cuyahoga County	J. L. Murphy	8.3	8.5	123	84	15	1	—	—	
Campbell Island	Fort Ancient Aspect, Upper Miss. Phase	ca. AD 1400-1600	Butler County	Brown & Anderson									"Eastern complex corn, beans, and sunflower seeds"
Reeves	Whittlesey Focus	ca. AD 1300-1400	Cuyahoga County	J. L. Murphy	11.0	6.8	4	—	75	—	25	—	Also seven large, crescent-shaped corn grains from 8-rowed ears.
McGraw	Hopewell phase, Scioto Tradition	ca. AD 450±	Ross County	O. F. Prufer	12.0	7.0	1	—	—	100	—	—	Also medium sized grains, not crescent shaped, not northern flint, not primitive pop. 1 kernel, 8-10-rowed, slightly crescent shaped.
Daines II Mound	Adena	C14 BC 280±140, U. of Mich.	Athens County	M. Moskal & J. L. Murphy	10.0	6.0	1	—	100	—	—	—	Tropical flint ear in husk, striated husk, tapered, cigar-shaped. Length 5.4 cm. Maxium diam. 2.2 cm.

OKLAHOMA

Site Name & No.	Culture/Tribe	Date	Location	Sample Furnished by	Mean Row No.	Median Cupule Width (mm)	No. Cobs	Row Numbers % of Total Cobs						Other Plant Remains/Comments
								8	10	12	14	16+		
Ferdinandina, Deer Cr., or Miller	Protohistoric French Contact Probably Wichita	ca. AD 1730-1760		B. Sudbury	10.0	8.8	1	—	100	—	—	—	Cupules fairly open, not narrow, but still eastern 8-rowed type cob. Corn grains from 8, 10, 12, and one from 14-rowed cob. 2 from 8-rowed wider than long, 1 from 12-rowed longer than wide.
Bryson (34KA5)	Early 18th Century, French Contact, probably Wichita	ca. AD 1700-1750		B. Sudbury	12.0	7.6	1	—	—	100	—	—	Grains were scraped off this cob when green. This large cob resembles some recent Cherokee corn in our collections (H. C. C.)
34HS24	Ft. Coffee Focus, Fulton Aspect	ca. AD 1500-1600		T. R. Cartledge	11.6	6.8	15	7	33	40	13	7	Also distorted corn grains, mostly from 12-rowed ears. Common bean, hickory nut.
Bowman (34LF66)	Culture?	ca. AD 1400-1500		C. J. Bareis & J. Brown	11.0	7.4	6	17	17	66	—	—	Also corn grains from 8, 10, and 12-rowed ears.
Geren (34LF36)	Culture?	ca. AD 1400-1500		J. Brown	12.0	5.1	3	33	—	33	—	33	Also corn grains from 10, 12, and 16-rowed ears, longer than wide.
Geren (34LF36)	Culture?	ca. AD 1400-1500		J. Brown	12.0	5.1	3	33	—	33	—	33	Also corn grains from 10, 12, and 16-rowed ears, longer than wide.
Van Schuyver (34PT20)	Washita R. Focus	ca. AD 1350-1550		R. E. Bell	11.7	6.3	12	8	33	33	17	8	Also corn grains from 10-rowed ears. Piñon nuts.
Stamper	Panhandle Aspect	C14 AD 1300±70 Wisc.-83	Texas County	C. J. Bareis	11.0	6.8	2	—	50	50	—	—	Also corn grains from 8, 10, and 12-rowed ears.
34KA131		ca. AD 1200-1350		B. Sudbury									Two small corn grains, one from 12-rowed ear, one probably popcorn. Wild sunflower, Iva.
Lee	Washita R. Focus	ca. AD 1200-1300	Garvin County	R. E. Bell	10.6	6.8	65	14	47	31	8	—	
Brackett	Gibson Aspect	C14 AD 1250±100	Cherokee County	C. J. Bareis	11.0	7.2	2	—	50	50	—	—	
Lee II (34GV4)	Washita R. Focus	ca. AD 1158		R. E. Bell	13.0	6.4	8	—	—	63	25	12	C. foetidissima.
Hughes	Gibson Aspect	C14 AD 1075±100	Muskogee County	C. J. Bareis	10.0	5.2	3	34	33	33	—	—	
Quartz Mt. Shelter	Culture?	ca. AD 1000+		T. Bastian	10.0	6.5	1	—	100	—	—	—	Not carbonized—may be modern, from disturbed deposits.

OKLAHOMA

Site Name & No.	Culture/Tribe	Date	Location	Sample Furnished by	Mean Row No.	Median Cupule Width (mm)	No. Cobs	8	10	12	14	16+	Other Plant Remains/Comments
Jones (34LF75)	Culture?	ca. AD 1000+		C. J. Bareis	8.0	9.5	1	100	—	—	—	—	Northern flint with shallow cupule. Also corn grains from 8, 10, and 12-rowed ears. Some are not crescent-shaped, probably from flint-like "chapalote".
Horton (34SQ11)	Culture?	ca. AD 1000+		C. J. Bareis	10.0	5.0	1		100	—	—	—	Hickory, butternut, <u>Viburnum prunifolium</u>.
Norman	Gibson Aspect	C14 AD 950±100	Wagoner County	C. J. Bareis									3 corn grains from 8-rowed ears and 1 from a 10-rowed.
Pruit	Early Woodland	ca. AD 600-800	Murray	Trufant Hall	10.7	6.5	3	—	67	33	—	—	
Kenton Cave 1 (34CIKE1)	Mixed, some possibly early Basketmaker	ca. AD 500		C. J. Bareis & R. E. Bell	10.5	6.3	50	34	18	38	10	—	Also grains of flint and popcorn; <u>Not carbonized</u>. <u>Lagenaria</u>, acorn, <u>C. foetidissima</u> and <u>Opuntia</u>.

PENNSYLVANIA

Site Name & No.	Culture/Tribe	Date	Location	Sample Furnished by	Mean Row No.	Median Cupule Width (mm)	No. Cobs	Row Numbers % of Total Cobs						Other Plant Remains/Comments
								8	10	12	14	16+		
Site 3 Miles S of Waynesburg	Delaware Proto-historic & Historic		Franklin County	P. R. Stewart										Small amount 8-rowed flint, small dia., typical of northern flint types. 4 fragments, perhaps pumpkin flesh replaced by mud, common bean, *Chenopodium* sp.
Sheeprock Shelter	Shenk's Ferry & Susquehanock	most ca. AD 1550 & after est.	Huntington County	Fred Kinsey	8.3	11.3	1714	88	11	1	—	—		Not carbonized. *C. pepo, Lagenaria*, common bean, peach, hickory nut, wild plum, black walnut, acorn, hackberry, sunflower, wild cherry, wild grape, butternut, chestnut, basswood, redbud
36AL40	Late Prehistoric Monongahela			W. Buker by R. L. George			←——————(1714 cobs)——————→				(42 cobs)			Approx. 50 corn grains, 7.2 to 11.5 mm wide, wider than long and crescent-shaped. Common beans, median length 10.5 mm, median width 6.0 mm, black walnut.
Parker (36LU14)	Iroquoian? Susquehanock?	ca. AD 1450-1500		I. F. Smith III										Numerous northern flint corn grains. Most are large, crescent-shaped and appear to be from 8-rowed ears. (See Pa. Arch. 48(3):29). Common bean, hickory nut, wild plum, persimmon, butternut.
McKees Rocks (36AL16)	A regional Focus of Ft. Ancient Aspect	ca. AD 1500		W. E. Buker	8.0	7.7	18	100	—	—		—		There is a report on the corn from this site by Volney Jones in *Pennsylvania Archaeologist*.
Murphy's Old (36AR129)	Most plant remains probably prehistoric Monongahela	ca. AD 1500		R. L. George	8.7	9	3	67	33		—	—		Hickory, hawthorn.
Drew (36AL62)	Monongahela	ca. AD 1000-1300		W. E. Buker										Approx. 5 qts. of carbonized corn grains, etc. Corn is all 8 and 10-rowed northern flint, with crescent-shaped grains wider than long. Most are over 8 mm wide and some are 13 mm wide. Hickory nut.
Ryan (36WM23)	Monongahela	ca. AD 1200		R. L. George	9.0	7.7	2	50	50	—		—		Common beans, hickory nut.
36BT43	Late Woodland-Monongahela			R. L. George										Hickory, acorn or hazelnut meat.

PENNSYLVANIA

Site Name & No.	Culture/Tribe	Date	Location	Sample Furnished by	Mean Row No.	Median Cupule Width (mm)	No. Cobs	Row Numbers % of Total Cobs					Other Plant Remains/Comments
								8	10	12	14	16+	
36BT43	Late Woodland-Monongahela			R. L. George									Hickory, acorn or hazelnut meat.
Gnagey (36SO55)	Late Woodland-Monongahela	C14 AD 880-1070±		R. L. George	8.4	8.2	51	82	14	5	—	—	Common beans, median length 9.6 mm, median width 5.4 mm, <u>C. pepo</u>, hickory, wild plum, acorn, hazelnut, butternut, chestnut, hawthorn, <u>Polygonum</u> sp., sedge, <u>Ilex</u> sp.

SOUTH CAROLINA

Site Name & No.	Culture/Tribe	Date	Location	Sample Furnished by	Mean Row No.	Median Cupule Width (mm)	No. Cobs	Row Numbers % of Total Cobs					Other Plant Remains/Comments
								8	10	12	14	16+	
Ft. Moore (38AK4-26)	Historic White or Indian	AD 1716-55 or AD 1680-1765		T. M. Ryan	8.8	11.4	8	62	38	—	—	—	
1670 Charlestowne (38CHL-61C)	Indian	pre AD 1670		T. M. Ryan	9.0	8.8	2	50	50	—	—	—	Hickory nut, persimmon, black gum
McCollum (38CS2)	Complicated Stamped pottery	ca. AD 1400-1600		T. M. Ryan									4 fragments of a 10-rowed cob. Greatest thickness 3.5, cob width 7.5 mm. Hickory nut, persimmon

SOUTH DAKOTA

Site Name & No.	Culture/Tribe	Date	Location	Sample Furnished by	Mean Row No.	Median Cupule Width (mm)	No. Cobs	Row Numbers % of Total Cobs					Other Plant Remains/Comments
								8	10	12	14	16+	
Ft. Pierre II (39ST217)	Historic			R. L. Stephenson									Wild plum, wild cherry, domestic peach
Four Bear (29DW2)	P-CC Late Arikara	Tree ring date AD 1758-74+		G. Agogino	8.3	9.9	20	85	15	—	—	—	Moderately large *C. pepo*, common bean, wild plum
Bonesteel	Late Arikara		Gregory County	G. Agogino	9.3	7.5	6	67	33	—	—	—	
Phillips Ranch (39ST14)	P-CC Snake Butte Focus, Bad R. phase, Protohistoric Arikara	AD 1750-1800		R. L. Stephenson	8.7	8.1	37	70	30	—	—	—	
Mush Creek (39HU5)	P-CC Probably Arikara	18th Cen. AD		R. B. Johnson									Unidentified grass seed.
Bowman (39HU204)	P-CC	18th Cen. AD		R. B. Johnson	9.0	9.0	21	52	43	5	—	—	Corn grains, mostly from 8 & 10 rowed cobs, crescent-shaped. *C. pepo*, common bean, wild plum.
Cheyenne R. Village (39ST1)	P-CC Protohistoric Arikara, prob. Bad R. phase	ca. AD 1700-1800		W. H. Over & R. L. Stephenson	8.8		10	90	—	—	—	10	*C. pepo*, wild plum, wild cherry, hackberry
Buffalo Pasture (39ST9)	P-CC Protohistoric Arikara, prob. Bad R. phase	ca. AD 1700-1800		W. H. Over	8.7		3	67	33	—	—	—	
39ST25	P-CC Protohistoric Arikara, prob. Bad R. phase	ca. AD 1700-1800											*C. pepo* and possibly var. oxifera.
Pascal Creek (39AR207)	P-CC Arikara affinities to Stanley or Snake Butte focus, Bad R.	18th Cen.		R. L. Stephenson									Two varieties of *C. pepo*, one of which cultivar "Mandan"

SOUTH DAKOTA

Site Name & No.	Culture/Tribe	Date	Location	Sample Furnished by	Mean Row No.	Median Cupule Width (mm)	No. Cobs	Row Numbers % of Total Cobs					Other Plant Remains/Comments
								8	10	12	14	16+	
Dodd, Upper level (39ST30)	P-CC Protohistoric Arikara, Bad R. phase	AD 1700-1750		R. L. Stephenson	8.0	10.1	6	100	—	—	—	—	—
Ft. George Village (29ST17)	P-CC Protohistoric Arikara, Bad R. phase	Early to Mid 18th Century		J. J. Hoffman	8.6	8.9	60	72	25	3	—	—	C. pepo, "Red Lodge" and "Mandan" types, common beans
Bamble (39CA6)	P-CC Protohistoric Arikara	AD 1690-1750		D. A. Baerreis	8.3	10.3	37	87	13	—	—	—	Approx. 100 C. pepo seeds 6.5 x 8 to 9.5 x 16.5 mm.
Spiry-Eklo (39WW3)	P-CC Protohistoric Arikara LeBeau phase	AD 1690-1750		D. A. Baerreis	8.8	9.2	89	67	25	8	—	—	C. pepo, small to medium.
Steamboat Creek (39PO1)	P-CC				8.0	8.0	1	100	—	—	—	—	—
Medicine Crow (39BF2)	P-CC			R. L. Stephenson	9.0		21	62	28	5	5	—	C. pepo, like "Mandan," common bean, sunflower
Sully (39SL4)	P-CC			R. L. Stephenson & C. H. McNutt	8.9		13	61	31	8	—	—	C. pepo, large and small, wild grapes, plum, Chenopodium.
39SL28	P-CC Stanley and Russell Ware			C. H. McNutt	8.0	9.0	1	100	—	—	—	—	—
Coleman Village (39SL3)	EC and P-CC Stanley in assoc. with historic material			R. L. Stephenson & C. H. McNutt	8.4		9	89	—	11	—	—	C. pepo
39SL13	Historic, Stanley, Russell & Riggs pottery			C. H. McNutt									C. pepo seeds up to 18.5 x 9.0 mm
Medicine Creek (39LM222)	EC Shannon Focus Chouteau Aspect	ca. AD 1650		R. L. Stephenson	8.0	5.7	1	100	—	—	—	—	Wild plum, juniper, Chenopodium

SOUTH DAKOTA

Site Name & No.	Culture/Tribe	Date	Location	Sample Furnished by	Mean Row No.	Median Cupule Width (mm)	No. Cobs	Row Numbers % of Total Cobs					Other Plant Remains/Comments
								8	10	12	14	16+	
Laroche (39ST9) Component A	EC. Proto Arikara-Pawnee or ancestral Pawnee	AD 1600-1695		J. J. Hoffman									Medium to large corn grain and 10-rowed cob with c. w. 7.6 mm. Not carbonized.
Laroche (39ST9)	Component uncertain	ca. AD 1450-1695		G. Agogino	9.0	7.8	23	57	39	4	—	—	
39LM224	EC. Shannon Focus, Chouteau Aspect	ca. AD 1650		R. L. Stephenson									Hackberry seeds.
Demery (39CO1)	EC. Prehistoric Mandan modified by Arikara	ca. AD 1600-1650		W. R. Wood	9.2	7.6	123	55	33	10	2	—	C. pepo, common bean, wild grape, juniper, pop, flint, flour corn grains.
Spain (39LM301)	EC. Chouteau Aspect	AD 1550-1650		C. S. Smith	8.9	9.5	7	57	43	—	—	—	Wild plum, hackberry
Strickler (39LM1C)	EC. Chouteau Aspect, similar to Spain	AD 1550-1650		C. S. Smith & D. A. Baerreis	8.0	7.6	1	100	—	—	—	—	Wild plum.
Spiry (39WW10)	EC. Akaska Focus	ca. AD 1600		D. A. Baerreis	9.3	8	3	67	33	—	—	—	
Rygh (39CA4)	EC.	ca. AD 1500-1600		R. L. Stephenson	8.0	10.4	4	100	—	—	—	—	C. pepo large and small, common beans.
Cattle Oiler, Upper component (39ST224)	EC.	ca. AD 1500-1600		D. E. Moerman & D. D. Jones	8.0	9.4	1	100	—	—	—	—	"Northern Flint", C. pepo, medium to large.
Good Soldier (39LM238)	EC? Shannon Focus, Chouteau Aspect	before AD 1600		R. L. Stephenson									Wild plum.
Hosterman (39PO7)	EC. Lecompte Focus, Chouteau Aspect	ca. AD 1550		R. L. Stephenson	8.5	8.7	159	79	18	3	—	—	Corn grains, mostly 8 and 10-rowed. C. pepo, small and medium and a few larger. Two are possibly C. mixta. One common bean, wild plum, Chenopodium?

SOUTH DAKOTA

Site Name & No.	Culture/Tribe	Date	Location	Sample Furnished by	Mean Row No.	Median Cupule Width (mm)	No. Cobs	8	10	12	14	16+	Other Plant Remains/Comments
39SL23	EC Russell Ware and high incidence of collared rims			C. H. McNutt	8.0	8.9	1	100	—	—	—	—	
Laroche (39ST9) Component B	IC?	C14 AD 1460-1580		J. J. Hoffman									*P. americana*, *P. hortulana*
39SL19	IC? Russell pottery			C. H. McNutt									*C. pepo* 19.2 x 10.4 mm
Black Partizan, Component B (39LM218)	IC	ca. AD 1450		R. L. Stephenson									12, 14, and 16-rowed cobs. Also 7 8-rowed and 14 10-rowed cobs analyzed by Nickerson. *C. pepo*, common bean, wild plum, *Chenopodium* sp.
Crow Creek (39BF11)	IC Mostly Campbell Creek Focus	ca. AD 1400		M. Kivett	10.1	8.3	76	34	32	31	8	—	*C. pepo* var. *ovifera*, also some like Conn. field pumpkin and Mandan, *Chenopodium* seeds.
John Ketchen (39ST223)	EMM.	ca. AD. 1200-1400		D. T. Jones									12, 14, 16-rowed cob fragments and grains. 1 12-rowed possibly sweet, others are flint and flour, grains longer than wide.
Sully School (39SL7)	EMM.			R. L. Stephenson									*Prunus americana*, *Prunus virginiana*.
Sommers (39ST56)	IMM.	Long occupation beginning end 13th cen.		R. E. Jensen	12.0	6.4	3	—	33	33	33	—	Bluestem grass seeds, *Cleome serrulata*.
39ST12	IMM	Tree ring date AD 1409+		R.L. Stephenson									8, 10, 12-rowed corn cob fragments. *C. pepo*, common beans, *P. americana*, *Ambrosia altemifolia*
Jiggs Thompson (39LM208)	IMM			R. L. Stephenson									12, 14, 16-rowed cobs.

SOUTH DAKOTA

Site Name & No.	Culture/Tribe	Date	Location	Sample Furnished by	Mean Row No.	Median Cupule Width (mm)	No. Cobs	8	10	Row Numbers % of Total Cobs 12	14	16+	Other Plant Remains/Comments
Prettyhead (39LM2, 39LM232)	IMM				8.0	7.6	1	100	—	—	—	—	
Akichita (39BF221)	IMM			R. L. Stephenson									Hip of *Rosa setigera*.
Langdeau (39LM209)	Grand Detour Phase, IMM			W. W. Caldwell									Seeds *Lithospermum carolinense*.
12 Mile Creek (39HT1)	Initial Middle Missouri	ca. AD 1000-1200		D. Benn									One corn shank with diameter of approx. 17 mm.
Cattle Oiler Lower component (39ST224)	IMM	C14 AD 1110		D. E. Moerman & D. T. Jones	12.0	4.2	2		—	100		—	"Small pop corn, open cupules". Basswood, hackberry.
Fay Tolton (39ST11)	IMM	C14 AD 1090±150		R. L. Stephenson									*C. pepo*, medium size. *P. americana*. *P. virginiana*.
Mitchell (39DV2)	Initial Middle Missouri, Over Focus	C14 AD 985-1125±65		D. Benn	11.0	9.5	2		50	50		—	Also cupules of very small 8 and 10-rowed popcorn with cupule widths of 3.5 to 5.5 mm. Also corn grains from 8 and 10-rowed ears; some longer than wide and others wider than long. Sunflower.
Arp (39BR101)	Loseke Creek Focus, Late Woodland	C14 AD 750±110		W. R. Hunt	12.0	5.0	1		—	100		—	
Grover Hand Mound (39DW240)	Middle Woodland Hopewell relationship	C14 AD 230±75		R. W. Neuman									Strips of inner bark.
Unnamed village near Pierre	Culture?	Time?	Hughes County	W. H. Over	8.0		3	100	—	—		—	

Archaeological Sites East of the Rockies 141

TENNESSEE

Site Name & No.	Culture/Tribe	Date	Location	Sample Furnished by	Mean Row No.	Median Cupule Width (mm)	No. Cobs	Row Numbers % of Total Cobs						Other Plant Remains/Comments
								8	10	12	14	16+		
Chucalissa State Park (40SY1)	Walls Phase, Middle Mississippian	ca. AD 1500-1600±		C. H. Nash	11.7	6.1	257	6	31	43	16	5		Common bean, hickory nut, persimmon, acorn, pecan, sunflower, butternut
Mound Bottom (40CH8)	Middle to Late Mississippian			M. O'Brien	10.9	6.4	30	13	37	44	3	3		
40LN86	Late Woodland - Mississippian	ca. AD 1200-1300		B. M. Butler										Crescent-shaped corn grains, 6.0 to 13.7 mm. wide, most from 8-rowed ears, one from 12-rowed, 8.0 mm wide. Hickory nut.
Sellar's (40WI1)	Mississippian	corrected C14 1025-1275		B. Butler	9.7	6.7	6	33	50	17	—	—		10 and 12-rowed corn grains 5.5 to 10 mm wide. Halves of common beans, 8.6 to 10.1 mm long.
Rankin Site (40CK6)	Early Woodland			C. H. McNutt	9.1	8.8	16	50	44	6	—	—		Hickory nut, grass seeds, passion flower (*P. incarnata*).
40FR8	Culture?	Time?		C. H. Faulkner										(1966) Faulkner, C. H. and Graham, J. B. Food Plant Remains on Tennessee Sites. Contains a listing of food plant remains from over 40 Tennessee sites.
40FR13 (Fea. 3)	Culture?	Time?		C. H. Faulkner										

142 *Archaeological Sites East of the Rockies*

TEXAS

Site Name & No.	Culture/Tribe	Date	Location	Sample Furnished by	Mean Row No.	Median Cupule Width (mm)	No. Cobs	8	10	12	14	16+	Other Plant Remains/Comments
Mission San Lorenzo de la Santa Cruz, Sp. Mission	Spanish Mission	AD 1762-1769		C. Tunnell									Approx. 196 corn grains, mainly from 14-16-18-rowed, long grained dents, similar to recent southern and some Mexican dents. A few from 8-10-rowed, usually shorter and broader. 20 *Opuntia* seeds, probably *O. engelmannii*.
Gilbert (41RA3)	Probably Wichita	Last half 1700s		E. R. Jelks	12.0	6.7	1	—	—	100	—	—	
San Juan Capistrano Mission (41BX5)	Coahuiltecan Indians	AD 1731-1762		Mardith Schuetz	14.5	7	98	—	1	19	45	35	
A-61 near border on Canadian R.	Antelope Cr. Focus, Panhandle Aspect	ca. AD 1250-1400	Hutchinson County	L. F. Duffield	11.3	4.4	3	—	33	67	—	—	5 corn grains from 10 and 12-rowed ears. 1 legume seed and 1 juniper seed
Spring Canyon (41HC20)	Antelope Cr. Focus, Panhandle Aspect	ca. AD 1250-1400		L. F. Duffield	14.0	7.0	1	—	—	—	100	—	Tip of woody plant, not corn.
Kyle Rock Shelter	Toyah Focus, Central Texas Aspect	ca. AD 1200-1500	McLennon County	E. R. Jelks	10.0	5.5	1	—	100	—	—	—	Moderately long glumes, deep flexible cupules. Not carbonized.
Near Amarillo	Assoc. with Archaic artifacts	No date available	Potter County	G. Agogino	10.0	5.6	1	—	100	—	—	—	Probably very flat hard-kernelled flint. Cob is short and slightly pyramidal with broadening at base.
Big Ben Area, Cave I	No Data			F. M. Setzler	9.3	?	3	33	67	—	—	—	Not carbonized.
Big Ben Area, Mule Ears Cave I	No Data			F. M. Setzler	12.0	?	1	—	—	100	—	—	Not carbonized.
Big Bend Area Sunny Glen Cave II	No Data			F. M. Setzler									In rawhide bag: Large common beans, white, deep purple and purple. One *C. pepo* seed 9 X 15.5. Five flint corn grains 5 x 9.5, 4.5 x 9, 5 x 9. Not carbonized.

VIRGINIA

Site Name & No.	Culture/Tribe	Date	Location	Sample Furnished by	Mean Row No.	Median Cupule Width (mm)	No. Cobs	Row Numbers % of Total Cobs				Other Plant Remains/Comments
								8	10	12	14	16+
Hand (44SN22)	Perhaps Nottaway Indian	AD 1500-1620		G. P. Smith	8.9	7.3	110	60	35	5	—	—	Also hickory, persimmon, pine cones

WEST VIRGINIA

Site Name & No.	Culture/Tribe	Date	Location	Sample Furnished by	Mean Row No.	Median Cupule Width (mm)	No. Cobs	Row Numbers % of Total Cobs				Other Plant Remains/Comments
								8	10	12	14	16+
Site in northern panhandle	Culture?	Late Prehistoric		R. C. Dunnell, U. of Kentucky									8 to 10-rowed corn grains. "A few of the grains are large, crescent-shaped like northeastern flint, some are medium-sized and a few of these may be 10-rowed. Embryos have burned out, leaving bean-like shapes." Black walnut.
Fairchance (46MR13)	Middle Woodland	C14 AD 160, 420.		E. T. Hemmings									Hickory, wild plum, acorn, Amaranthus sp., butternut, honey locust, sumac, ground cherry.

WISCONSIN

Site Name & No.	Culture/Tribe	Date	Location	Sample Furnished by	Mean Row No.	Median Cupule Width (mm)	No. Cobs	8	10	12	14	16+	Other Plant Remains/Comments
Crabapple (47JE93)	Historic Winnebago	ca. AD 1775-1825		J. Spector	8.3	8.2	211	88	11	1	—	—	
Marina, Madaline Island, Lake Superior, Wis.	Probable historic Chippewa, Euro-Indian	18th Century		R. A. Birmingham	8.5	7.2	55	82	13	3	2	—	One example of pop corn
Bell (47WN9)	Historic Fox	Early AD 1700s		W. L. Wittry	8	6.4	3	100	—	—	—	—	*C. pepo*, *Prunus americana*, basswood matting
47GL122	Late Woodland & Middle Mississippian	AD 1400-1500		W. L. Wittry	10.7	5.8	3	—	67	33	—	—	"Open cupules." Also two broken corn grains. *Scirpus* sp. and uncarbonized wild seeds that are probably modern.
Midway (47LA19)	Oneota Upper Mississippian	C14 AD 1420±70 (WISC. 61)		R. Peske									Seven corn grains distorted by charring, from 8 and 10-rowed ears. Yarnell has reported common beans from this site.
Bornick	Grand R. Focus, aberrant Oneota	ca. AD 1290		G. Gibbon									Corn grains from 8 and 10 and possibly a 12-rowed ear. Hickory nut, hazelnut.
McCauley Campsite	Upper Mississippian		Fond du Lac County	D. A. Baerreis	8.2	8.2	12	92	8	—	—	—	"Wide short cupules. Northern flint. Cob firm."
Dietz	Late Woodland	ca. AD 1250	Dane County	D. A. Baerreis	12.0	6.5	1	—	—	100	—	—	
Walker-Hooper (47GL65)	Grand R. Focus, aberrant Oneota	ca. AD 1200-1250		G. Gibbon	corn cobs 9.9 corn grains 8.8	6.0	19 66	53 68	21 24	10 8	16 —	— —	*C. pepo*, Common bean, hickory nut, hazelnut
Aztalan (47JE1)	Middle Mississippian	ca. AD 1100-1300		D. A. Baerreis & J. E. Freeman	12.0	6.0	1	—	—	100	—	—	Also mass of charred corn grains, mostly 12-rowed. "Rounded tops, not crescent-shaped, probably small flint."

Row Numbers % of Total Cobs (columns 8, 10, 12, 14, 16+)

WISCONSIN

Site Name & No.	Culture/Tribe	Date	Location	Sample Furnished by	Mean Row No.	Median Cupule Width (mm)	No. Cobs	Row Numbers % of Total Cobs					Other Plant Remains/Comments
								8	10	12	14	16+	
Lasley's Point (47WN8, 47WN96)	Oneota, Upper Mississippian	ca. AD 1000-1250		R. Peske									Fifteen corn grains distorted by charring, mostly 8-10 rowed, but including three 10 or 12-rowed and one from 14-rowed ear.
Pipe (47FD10)	Grand River phase Oneota	ca. AD 1200		D. F. Overstreet	9.3	7.6	3	67	—	33	—	—	Possible common bean, hickory, wild plum, butternut
Diamond Bluff (47PI2)	Late Woodland, Oneota or Middle Mississippian	ca. AD 1150±		R. A. Alex	9.7	6.9	36	39	41	17	3	—	Wild plum
Carcajou Point (47JE2)	Oneota, Upper Mississippian	ca. AD 1000		R. Peske	10.7	7.0	30	7	53	37	3	—	Also 10 corn grains, small to medium size, mostly slightly crescent-shaped.
Richter	Northern Tier Middle Woodland	ca. AD 0-200	Door County	G. R. Peters									Acorn, butternut, unidentified objects, some probably legumes.

ONTARIO, CANADA

Site Name & No.	Culture/Tribe	Date	Location	Sample Furnished by	Mean Row No.	Median Cupule Width (mm)	No. Cobs	Row Numbers % of Total Cobs					Other Plant Remains/Comments
								8	10	12	14	16+	
Midland	Huron	AD 1649		M. Greenwald	8	9.3	4	100	—	—	—	—	
BDGU-1-122	Protohistoric Huron Iroquois	AD 1570-1590		W. C. Nobel	8.4	7.8	9	78	22	—	—	—	Also crescent-shaped, distorted corn grains, all or most from 8-rowed ears.
McIvor (BF FU-1)	Prehistoric Onondaga	ca. AD 1500		J. V. Wright	9.0	8.2	37	68	16	16	—	—	Also carbonized corn grains, mostly from 8-rowed ears but some from 10 and 12-rowed; all crescent-shaped.

ONTARIO, CANADA

Site Name & No.	Culture/Tribe	Date	Location	Sample Furnished by	Mean Row No.	Median Cupule Width (mm)	No. Cobs	Row Numbers % of Total Cobs					Other Plant Remains/Comments
								8	10	12	14	16+	
Draper	Iroquois	ca. AD 1450		D. M. Stothers	8.9	7.5	9	56	44	—	—	—	Also approximately 20 swollen corn grains. Five that were measurable were from 8-rowed ears and 7.7 to 11.0 mm wide. Wild plum, probably P. nigra or P. americana.
Bennett (AI GX-1)	Iroquois	ca. AD 1260		J. V. Wright									Crescent-shaped, distorted corn grains, all from 8-rowed ears; widths 7.7 to 11.0 mm.
Dewaele (AFHD-1)	Glen Meyer, Early Ontario Iroquoian	C14 1050-1095±90		W. A. Fox	8.9	8.0	25	68	20	12	—	—	Corn cob cupules compressed, generally well thickened, except in case of 12-rowed, which had semi-open cupules. C. pepo seed 11.3 x 6.8 mm, black walnut or butternut, acorn, juniper, hawthorn
Porteous (AGHB-1)	Terminal phase, Grand R. Focus, Princess Point	C14 820±100		D. M. Stothers	8.0	7.3	1	100	—	—	—	—	Cupules open, cob hard and thickened. Also swollen corn grains, 6.9 to 10.7 mm wide. Hazelnut.
Grand Banks		ca. AD 600		D. M. Stothers									1 corn grain from 8-rowed ear, 9 mm wide, 7 mm long.

11 Published Works of Cutler and Blake

This is a complete listing of all works published by Hugh Cutler and Leonard Blake as of November 2000.

Anderson, Edgar, and Hugh C. Cutler
- 1942 Races of *Zea mays:* I. Their Recognition and Classification. *Annals of the Missouri Botanical Garden* (St. Louis) 29:69–88.
- 1950 Methods of Corn Popping and Their Historical Significance. *Southwestern Journal of Anthropology* 6:303–8.

Blake, Leonard W.
- 1962 Analysis of Vegetal Remains from the Lawhorn Site. *Missouri Archaeologist* (Columbia) 24:97–98.
- 1974 Corn in the Province of Aminoya. *Quarterly Newsletter of Illinois Association for the Advancement of Archaeology* 6(1): 15.
- 1981 Carbonized Plant Remains from the Bonnie Brook Site (36Bt43) (Appendix A). *Pennsylvania Archaeologist* 51:52–53.
- 1981 Early Acceptance of Watermelon by Indians of the United States. *Journal of Ethnobiology* 1(2): 193–99.
- 1981 Floral Remains from the 1978–1979 Street Excavations at Michilimackinac. Appendix in *Excavations at Fort Michilimackinac: 1978–1979. The Rue de la Babillarde*, by Donald P. Heldman and Roger T. Grange. Archaeological Completion Report Series No. 3, Mackinac Island State Park Commission, Mackinac Island, Michigan.
- 1985 Cultivated Plant Remains from Historic Missouri and Osage Indian Sites. Chapter 6.6 in *Osage and Missouri Indian Life Cultural Change: 1675–1825*, pp. 832–47. Final Performance Report on National Endowment for the Humanities Research Grant RS-20296, Dec. 31, 1985.
- 1986 Corn Analysis. In *The Bonnie Creek Site: A Late Mississippian Homestead on the Upper Balum Creek Valley, Perry County, Illinois*, by Mark Wagner, pp. 273–77. Illinois Historic Preservation Series No. 3, American Resources Group, Ltd., Carbondale.
- 1986 Corn and Other Plants from Prehistory into History in Eastern United States. In *The Protohistoric Period in the Mid-South, 1500–1700*, edited by David H. Dye and Ronald C. Brister, pp. 3–13. Archaeological Report No. 18, Mississippi Department of Archives and History, Jackson.
- 1987 A Prehistoric Indian's Menu. In *Discovery* (St. Louis Science Center, St. Louis) 7(3): 8–10.

1988 Carbonized Plant Remains from the Burris Site (3Cg18). In *The Burris Site and Beyond,* edited by Marvin Jeter, pp. 140–46. Arkansas Archaeological Research Series No. 27, Fayetteville.

1988 Food Plants of the Historic Osage Indians. *Missouri Folklore Society Journal* 10:15–21.

1988 Plant Remains from the Standridge (3Mn53) and Amos (3Mn622) Sites. In *Standridge,* by Ann M. Early, pp. 133–36. Arkansas Archaeological Survey Research Series No. 29, Fayetteville.

1989 Comments on Some Native Food Plants Used by the Osage Indians. *Nature Notes: The Journal of the Webster Groves Nature Study Society* (St. Louis) 61(1): 1–3.

1989 Review of *Prehistoric Agriculture in the Central Plains,* by Mary Adair, Publications in Anthropology 16, Department of Anthropology, University of Kansas, Lawrence. *Missouri Archaeological Society Quarterly* (Columbia) 2:16–18.

1990 Floral Remains. In *The Archaeology of the Cahokia Palisade. Part 2, East Palisade Investigations,* pp. 135–45. Illinois Culture and Resources Study No. 14, Illinois Historic Preservation Agency, Springfield. (Note: Identification of beans is incorrect, LWB.)

1992 Corn from the Orchard Site (22Le515). *Mississippi Archaeology* (Mississippi Department of Archives and History, Jackson) 7(1): 60–71.

1993 Pawpaws. *Missouri Archaeological Society Quarterly* (Columbia) 10(20): 20.

Blake, L. W., and Hugh C. Cutler

1963 Plant Materials from the Bell Site (Wn 9). *Wisconsin Archaeologist* (Madison) 44(1): 70–71.

1970 The Builders of Cahokia Mounds and Their Cultivated Plants. *Missouri Botanical Garden Bulletin* 58:25–32.

1972 Osage Indian Watermelons. *Missouri Archaeological Society Newsletter* (Columbia) 258:2.

1972 Plants from the Marty Coolidge Site, Illinois. In *The Marty Coolidge Site, Monroe County, Illinois,* by L. Carl Kuttruff, pp. 106–8. Southern Illinois Studies 10, University Museum, Southern Illinois University, Carbondale.

1974 Identification of Charred Organic Material Recovered from the Knappenberger Test Excavation (3Ms53). *Arkansas Archaeologist* (Fayetteville) 15:72.

1975 Floral Remains from the Hood Site (11Mv56). In *The Hood Site: A Late Woodland Hamlet in the Sangamon Valley of Central Illinois,* by R. Barry Lewis, pp. 8, 28–29. Illinois State Museum Reports of Investigations No. 31, Springfield.

1975 Food Plant Remains from the Zimmerman Site (11Ls13). Appendix 2 in *The Zimmerman Site,* by Margaret Kimball Brown, pp. 92–94. Illinois State Museum Reports of Investigations No. 32, Springfield.

1976 Floral Remains from House One of the South Southeast Row House Excavation of 1977. Appendix 4 in *Excavations at Fort Michilimackinac 1976: House One of the South Southeast Row House,* by Donald P. Heldman. Archaeological Completion Report Series No. 2, Mackinac Island State Park Commission, Mackinac Island, Michigan.

1976 Floral Remains from the Southeast Row House Excavation of 1976. Appendix 4 in *Excavations at Fort Michilimackinac, 1976: The Southeast and South Southeastern Row Houses,* by Donald P. Heldman. Archaeological Completion Report Series No. 1, Mackinac Island State Park Commission, Mackinac Island, Michigan.

1979   Plant Remains from the Upper Nodena Site (3Ms4). *Bulletin of the Arkansas Archaeological Society* 20:53–58.

1982   Plant Remains from the King Hill Site (23Bn1) and Comparisons with Those from the Utz Site (23Sa2). *Missouri Archaeologist* (Columbia) 43:86–110.

1983   Plant Remains from the Gnagey Site (36So55). Appendix 2 in *The Gnagey Site and the Monongahela Occupation of the Somerset Plateau,* by R. L. George. *Pennsylvania Archaeologist* 53(4): 83–88.

1986   Corn Analysis. In *The Kayenta Anasazi,* by S. Stebbins, B. Harrill, W. D. Wade, M. V. Gallagher, H. Cutler, and L. Blake, pp. 88–103. MNA Research Paper 30, Museum of Northern Arizona, Flagstaff.

Blake, Leonard W., and Rosalind M. Dean

1963   Corn from Plum Island. Appendix 3 in *The Plum Island Site,* by Gloria Fenner. Reports on Illinois Prehistory 1, Illinois Archaeological Survey Bulletin No. 4, University of Illinois, Urbana.

Bohrer, Vorsila L., Hugh C. Cutler, and Jonathan D. Sauer

1969   Carbonized Plant Remains from Two Hohokam Sites, Arizona BB 13–41 and Arizona BB 13–50. *Kiva* 35(1): 1–10.

Brooks, Richard H., Lawrence Kaplan, Hugh C. Cutler, and Thomas W. Whitaker

1962   Plant Material from a Cave on the Rio Zape, Durango, Mexico. *American Antiquity* 27(3): 356–69.

Cardenas, Martin, and Hugh Cutler

1943   Dos Variaciones Interesantes en el Mais. *Revista de Agricultura, University Simon Bolivar.* Cochabamba, Bolivia.

Carter (Cutler), Hugh, Leonard Blake, John Bower, and Winton Meyer

1964   *Vegetal Remains, Hosterman Site (39P07), South Dakota.* Bureau of American Ethnology Bulletin 189, pp. 231–32, Smithsonian Institution, Washington, D.C.

Cutler, Hugh C.

1939   Monograph of the North American Species of the Genus *Ephedra. Annals of the Missouri Botanical Garden* (St. Louis) 26:373–428.

1940   Plant Hunting to the Rainbow Bridge. *Yearbook 1940, American Rock Garden Society* (New York), pp. 25–27.

1940   A Rapid Method of Recording Angles. *American Midland Naturalist* 24:497–98.

1944   Medicine Men and the Preservation of a Relict Gene in Maize. *Journal of Heredity* (Washington, D.C.) 35(10): 290–94.

1945   Espiguetas de Dois Graos No Milho. *Anais Escola Superior de Agricultura "Luis Queiroz"* (Piracicaba, Brazil) 2:423–30.

1946   Exploration for Prehistoric Indian Foods. *Missouri Botanical Garden Bulletin* 34(8): 205–7.

1946   Races of Maize in South America. Botanical Museum (Harvard) Leaflets 12(8): 257–91.

1946   Rubber Production in Ceara, Brazil. Botanical Museum (Harvard) Leaflets 12(9): 301–16.

1947   A Comparative Study of *Tripsacum australe* and Its Relatives. *Lloydia* (Cincinnati) 10:229–34.

1947   Species Relations in *Cucurbita.* Abstract in *American Journal of Botany* 34:606.

1949   Razas de Maiz en Sud America (translation into Spanish by Martin Cardenas).

Revista de Agricultura de la Universidad Mayor de "San Simon" (Cochabamaba, Bolivia) 4-5:1–30.

1951 The Geographic Origin of Maize. *Chronica Botanica* 12:167–69.

1952 A Preliminary Survey of Plant Remains of Tularosa Cave. In *Mogollon Cultural Continuity and Change: Stratigraphic Analysis of Tularosa and Cordova Caves*, edited by P. S. Martin, J. B. Rinaldo, E. Bluhm, H. C. Cutler, and R. Grange, Jr., pp. 461–79. *Fieldiana: Anthropology* (Chicago) 40:1–529.

1953 Did Man Once Live by Beer Alone? Untitled comment by Hugh C. Cutler, p. 522, in a symposium edited by R. J. Braidwood, *American Anthropologist* 55(4): 515–26.

1953 Review of P. K. Reynolds' 1951 "Earliest Evidence of Banana Culture," Supplement No. 12 of the Journal of the American Oriental Society, Baltimore, Maryland, Dec. 1951. *American Anthropologist* 55(3): 443.

1954 Food Sources in the New World. *Agricultural History* 28:43–49.

1956 Corn from the Dietz Site, Dane County, Wisconsin. *Wisconsin Archaeologist* (Madison) 37(1): 18–19.

1956 The Plant Remains. In *Higgins Flat Pueblo, Western New Mexico*, edited by P. S. Martin, J. B. Rinaldo, E. A. Bluhm, and H. C. Cutler. *Fieldiana: Anthropology* (Chicago) 45:174–83.

1956 Vegetal Material from the Site of San Cayetanao. In *The Upper Pima of San Cayetano Del Tumacacori*, by C. C. DiPeso, pp. 459–60. Amerind Foundation Publication No. 7, Dragoon, Arizona.

1957 Comments on Prehistoric Corn Samples. In article by Charles Bareis. *Oklahoma Anthropological Society Newsletter* 6(5): 7–8.

1957 Corn from Reeve Ruin, Arizona. In *The Reeve Ruin of Southeastern Arizona*, by C. C. DiPeso, p. 122. Amerind Foundation Publication No. 8, Dragoon, Arizona.

1957 *The Identification of Non-Artifactual Archaeological Materials*, edited by Walter W. Taylor. National Research Council Publication 565, Report of conference held in Chicago, National Academy of Science, Washington, D.C.

1957 Report on Corn. In *Heltagito Rockshelter (NA 6380)*, by D. A. Breternitz. *Plateau* (Flagstaff, Arizona) 30(1): 1–16.

1958 Corn Cob from Shelter, Van Buren County, Arkansas. *Missouri Archaeological Society Newsletter* (Columbia) 119:4.

1959 Plant Materials from Six Oklahoma Sites. *Oklahoma Anthropological Society Newsletter* (Oklahoma City) 8(3): 4–7.

1960 Cultivated Plant Remains from Waterfall Cave, Chihuahua. *American Antiquity* 26:277–79.

1961 Measurements and Identification of Corn. Page 77 in *Fort Walton Mound Site in Houston County, Alabama*, by R. W. Neuman. *Florida Anthropologist* 14(3–4):75–80.

1961 Vegetal Remains: Corn and Cucurbits. In *A Survey and Excavation of Caves in Hidalgo County, New Mexico*, by Marjorie F. Lambert and J. Richard Ambler, pp. 90–94. School of American Research Monograph No. 25, Santa Fe.

1962 Agricultural Plant Remains, Kyle Site, Texas. Appendix 3 in *The Kyle Site, Hill County, Texas*, by E. B. Jelks, pp. 113–15. University of Texas Archaeology Series No. 5, Austin.

1962 Food Sources in the New World. In *Readings in Cultural Geography*, edited by P. L. Wagner and M. W. Mikesell, pp. 282–89. University of Chicago Press, Chicago.

1962   Review of *The Ethnobotany of Pre-Columbian Peru*, by Margaret A. Towle. *American Antiquity* 28(2): 256–57.

1963   Identification of Plant Remains. In *Second Annual Report, American Bottoms Archaeology*, edited by M. L. Fowler, pp. 16–19. Illinois Archaeological Survey, Urbana.

1963   Maize from a Basket Maker III Pithouse. In *Basket Maker III Sites near Durango, Colorado*, by Roy E. Carlson, pp. 46–47. University of Colorado Studies, Anthropology No. 8, Boulder.

1963   Plant Remains from a Grand Canyon Rockshelter Granary. *American Antiquity* 28(3): 301.

1964   Plant Materials from Four Cummings Mesa Sites. In *Survey and Excavations on Cummings Mesa, Arizona and Utah, 1960–1961*, by J. Richard Ambler, p. 102. Museum of Northern Arizona Bulletin 39, Flagstaff.

1964   Plant Remains from the Carter Ranch Site. *Fieldiana: Anthropology* (Chicago) 55:227–34.

1964   Review of Shreve and Wiggins' *Vegetation and Flora of the Sonoran Desert*, vols. 1 and 2. *Quarterly Review of Biology* 39(3): 296.

1965   Book Review: *Essays on Crop Evolution*, by Sir Joseph Hutchinson. *Science* 148:1711.

1965   Book Review: *Land and Life: A Selection from the Writings of Carl Ortwin Sauer*, edited by John Leighly. *Economic Botany* 19:89.

1965   Cultivated Plants. In *The McGraw Site*, by Olaf F. Prufer. *Scientific Publications of the Cleveland Museum of Natural History*, new series 4(1): 107–12.

1965   Identification of Plant Remains Excavated by the University of Illinois at the Cahokia Site. *Third Annual Report, American Bottoms Archaeology*, edited by M. L. Fowler, p. 10. Illinois Archaeological Survey, Urbana.

1965   Prehistoric Plant Remains from the Gila Bend Area. In *Salvage Archaeology in the Painted Rocks Reservoir, Southern Arizona*, by William W. Wasley and Alfred E. Johnson, pp. 108–9. University of Arizona Anthropological Papers 9, Tucson.

1966   *Corn, Cucurbits and Cotton from Glen Canyon.* Anthropological Papers No. 80, pp. 1–62, Department of Anthropology, University of Utah, Salt Lake City.

1966   Plant Materials from Canyonlands National Park, Utah. In *An Archaeological Survey of Canyonlands National Park*, by Floyd W. Sharrock, pp. 68–70. University of Utah Anthropological Papers No. 83, Salt Lake City.

1966   Plant Remains from the Fanning Site (14Dp1), Kansas. *Plains Anthropologist* 11(33): 210.

1966   *Plant Remains from the Grand Village of the Natchez.* Anthropological Papers of the American Museum of Natural History, vol. 51, pt. 1, p. 102, New York.

1966   Review of *Aboriginal Relationships between Culture and Plant Life in the Upper Great Lakes Region*. *American Antiquity* 31(4): 588–89.

1966   Review of *Essays on Crop Plant Evolution*, edited by Sir Joseph Hutchinson, Cambridge University Press, 1965. *American Antiquity* 31(5): 763–64.

1966   Review of *Plants and Civilization*, by H. C. Baker. *American Antiquity* 31(5): 763.

1967   Corn and Squash from Six Sites in North and South Dakota. In *Interpretation of Mandan Culture History*, by W. Raymond Wood. Bureau of American Ethnology Bulletin 198, pp. 177–82, Smithsonian Institution, Washington, D.C.

1967   Cultivated Plants from Picuris. (Previously unpublished manuscript. Published in this volume.)

1968 Missouri Botanical Garden's Fourteenth Systematics Symposium. *Bioscience* 18:134–35.

1968 The North American Pumpkin. *Missouri Botanical Garden Bulletin* 56(5): 5–6.

1968 Origins of Agriculture in the Americas. *Latin American Research Review* 3(4): 3–21.

1969 Corn and Cucurbits from Turkey Cave, NA2520 (Arizona). In *Archaeological Investigations in Turkey Cave (NA2520), Navajo National Monument, 1963*, by David A. Breternitz, pp. 22–26. Museum of Northern Arizona Technical Series No. 8, Flagstaff.

1969 Plant Remains from Sites near Navajo Mountain. In *Survey and Excavations North and East of Navajo Mountain, Utah, 1959–1962*, edited by Alexander J. Lindsay, pp. 371–78. Museum of Northern Arizona Bulletin 45, Flagstaff.

1969 Review of H. L. Edlin's *Man and Plants*. *Economic Botany* 23:292–93.

1970 Corn from Hogup Cave, a Fremont Site. Appendix 7 of *Hogup Cave*, by C. M. Aikens, pp. 271–72. University of Utah Anthropological Papers 93, Salt Lake City.

1970 Review of *The Domestication and Exploitation of Plants and Animals*, edited by Peter J. Ucko and G. W. Dimbleby. *American Scientist* (New Haven) 242(2): 216.

1970 Review of *Manual de Plantas Economical de Bolivia*, by Martin Cardenas. *Economic Botany* 24(1): 107–8.

1970 Review of *The Tepehuan of Chihuahua: Their Material Culture*, by C. W. Pennington. *Economic Botany* 24:228–29.

1971 Review of *Gardens, Plants and Man*, by Carlton B. Lees. *Missouri Botanical Garden Bulletin* 59(6): 20.

1971 Review of *Plant Agriculture*, edited by Jules Janick, R. W. Schery, F. W. Woods, and V. W. Ruttan. *Economic Botany* 25:340–41.

1972 High Protein Corn in the Garden's Collections. *Missouri Botanical Garden Bulletin* 60(1): 15–18.

1972 Living Off the Land. *Missouri Botanical Garden Bulletin* 60(2): 10–11.

1974 Review of *Seed to Civilization: The Story of Man's Food*, by Charles Heiser, Jr. *Quarterly Review of Biology* 49:279–80.

1975 Two Kinds of Gourds from Key Marco. In *The Material Culture of Key Marco*, by Marion Apjut Gilliloud, pp. 255–56. University of Florida Press, Gainesville.

1977 Flora. In *Prehistoric Ecology at Patarata 52, Veracruz Mexico: Adaptation to the Mangrove Swamp*, by Barbara L. Stark, pp. 179–81. Vanderbilt Publications in Anthropology No. 18, Nashville.

1977 Forward to *Uses of Plants by the Indians of the Missouri River Region*, by Melvin Gilmore, pp. v–xi. University of Nebraska Press, Lincoln.

1980 Corn from 38Bu162A. Appendix A in *The Discovery of Santa Elena*, by Stanley South, pp. 91–95. Research Manuscript 165, Institute of Archaeology and Anthropology, University of South Carolina, Columbia.

Cutler, Hugh C., and G. A. Agogino

1960 Analysis of Maize from the Four Bear Site and Two Other Arikara Locations in South Dakota. *Southwestern Journal of Anthropology* 16(3): 312–16.

Cutler, Hugh C., and Edgar Anderson

1941 A Preliminary Survey of the Genus *Tripsacum*. *Annals of the Missouri Botanical Garden* (St. Louis) 26(3): 249–69.

Cutler, Hugh C., and Leonard W. Blake

1966 Analysis of Banks Site, Arkansas, Corn. In *The Banks Village Site, Crittenden*

County, Arkansas, by Gregory Perino, pp. 146–50. Missouri Archaeological Society Memoir No. 4, Columbia.

1967 Notes on Plants from Sheep Rock Shelter. In *Archaeological Investigations of Sheep Rock Shelter, Huntington County, Pennsylvania,* vol. 1, by Joseph W. Michels and Ira F. Smith, pp. 125–46. Pennsylvania State University, University Park.

1968 Plants of the King's Hill Site (St. Joseph, Missouri). *St. Joseph Museum Graphic* 20(4): 4–9.

1969 Corn. In *Two House Sites in the Central Plains, an Experiment in Archaeology,* edited by Raymond Wood, pp. 61–62. Plains Anthropologist Memoir 6, vol. 14(44), pt. 2 (site 25Ft35).

1969 Corn from Cahokia Sites. *Illinois Archaeological Survey Bulletin* 7:122–36.

1969 Corn from San Juan Capistrano (Texas). State Building Commission Archaeological Program, Report 11, vol. 2, pp. 107–9.

1969 Floral Analysis: Friend and Foe Site (23Cl113). In *Doniphan Phase Origins: An Hypothesis Resulting from Archaeological Investigations in the Smithville Reservoir Area, Missouri,* by F. A. Calbrese, pp. 155–59. National Park Service Contract 14-10-2:920-50, University of Missouri, Columbia.

1970 Floral Remains from the Tyler-Rose Site, Oklahoma. In *The Tyler-Rose Site and Late Prehistory in East-Central Oklahoma,* by Thomas R. Cartledge, pp. 57–59. Oklahoma River Basin Survey, Archaeological Site Report 19, Norman.

1970 Food Plant Remains from Nine Prehistoric Indian Sites in the Yazoo Delta Area of Mississippi. *Mississippi Archaeological Association* (Clarksdale) 5(2): 1–6.

1970 Plants from Arizona Jt6:1. *Plateau* (Flagstaff, Arizona) 43:42–44.

1971 Corn from the South Park and Reeves Sites, Whittlesey Focus of Ohio. *Ohio Archaeologist* (Columbus) 21(3): 23–24.

1971 Floral Remains from the Knoll Spring Site, Cook Co., Illinois. *Illinois Archaeological Survey Bulletin* (Urbana) 8:244–46.

1971 Travels of Corn and Squashes. In *Man Across the Sea,* edited by C. L. Riley, J. C. Kelly, C. W. Pennington, and R. L. Rands, pp. 366–75. University of Texas Press, Austin.

1973 Corn from the Median Village Site (Utah). In *Median Village and Fremont Culture Regional Variation,* by John P. Marwitt, pp. 163–64. University of Utah Anthropological Papers 95, Salt Lake City.

1973 Plant Materials from the Kane Village Site (11Ms194), Madison County, Illinois. *Illinois Archaeological Survey Bulletin* (Urbana) 9:53–57.

1974 Plant Remains from the Callahan-Thompson Site (23Mi-71) A.D. 1400–1600. In *Mississippian Exploitative Strategies: A Southeast Missouri Example,* by R. Barry Lewis, pp. 62–63. Missouri Archaeological Society Research Series 11, Columbia.

1975 Plant Remains from the Fortified Hill Site (Arizona 7:13:8). *Kiva* 40:268–70.

1976 Carbonized Corn Remains from the Excavations. In *Papers on the Archaeology of Black Mesa, Arizona,* by George J. Gummerman and Robert C. Euler, pp. 106–7. Southern Illinois University Press, Carbondale.

1976 Corn from Snaketown. Appendix 4 in *The Hohokam,* by Emil Haury, pp. 365–66. University of Arizona Press, Tucson.

1976 *Plants from Archaeological Sites East of the Rockies.* Missouri Archaeological Society, University of Missouri, Columbia. Printed in microfiche, L. C. No. 76-49422. (A corrected and updated version of the 1973 edition. Published in this volume.)

1977 Plant Materials. Appendix D in *The Talking Crow Site,* by Carlyle S. Smith, pp. 185–87. University of Kansas Publications in Anthropology No. 9, Lawrence.

1980 Corn and Cucurbits. Appendix A in *Archaeological Explorations in Caves of the Point of Pines Region, Arizona,* by James C. Gifford. University of Arizona Anthropological Papers 36, Tucson.

1980 Plant Materials from Grand Canyon Sites. Appendix B in *Archaeology of the Grand Canyon: Unkar Delta,* by Douglas W. Schwartz, Richard C. Chapman, and Jane Kapp. School of American Research Press, Grand Canyon Archaeological Series, vol. 2.

1981 Corn and Squash from the Meadowcroft Rockshelter. Excerpts published in *The Appearance of Cultigens in the Upper Ohio Valley: A View from the Meadowcroft Rockshelter,* by J. M. Adovasio and W. C. Johnson. *Pennsylvania Archaeologist* 51:72–74.

Cutler, Hugh C., and John W. Bower

1961 Plant Materials from Several Glen Canyon Sites. In *Survey and Excavations in Lower Glen Canyon, 1952–1958,* by William Y. Adams, Alexander J. Lindsay, Jr., and Christy G. Turner II, pp. 58–61. Museum of Northern Arizona Bulletin 36, Glen Canyon Series No. 3, Flagstaff.

1967 Plant Material from Lower Glen Canyon Sites. In *Archaeological Investigations in Lower Glen Canyon, Utah, 1959–1960,* by Paul V. Long, pp. 69–70. Museum of Northern Arizona Bulletin No. 42, Glen Canyon Series No. 7, Flagstaff.

Cutler, Hugh C., and Martin Cardenas

1947 *Chicha, a Native South American Beer.* Botanical Museum (Harvard) Leaflets 1–3(3): 3360.

Cutler, Hugh C., and Marian C. Cutler

1948 Studies on the Structure of the Maize Plant. *Annals of the Missouri Botanical Garden* (St. Louis) 35:301–16.

Cutler, Hugh C., and Marcia Eikmeier

1967 Corn and Other Plant Remains. In *Clanton Draw and Box Canyon: An Interim Report on Two Prehistoric Sites in Hidalgo County, New Mexico and Related Surveys,* by Eugene B. McCluney, pp. 48–54. School of American Research Monograph No. 26, Santa Fe.

Cutler, Hugh C., and Lawrence Kaplan

1956 Some Plant Remains from Montezuma Castle and Nearby Caves. *Plateau* (Flagstaff, Arizona) 28(4): 98–100.

Cutler, H. C., and Winton Meyer

1965 Corn and Cucurbits from Wetherill Mesa. *American Antiquity* 31:136–52.

Cutler, Hugh C., and T. W. Whitaker

1956 *Cucurbita mixta* Pang: Its Classification and Relationships. *Bulletin of the Torrey Botanical Club* 83(4): 253–60.

1961 History and Distribution of the Cultivated Cucurbits in the Americas. *American Antiquity* 26(4): 469–85.

1967 Cucurbits from Tehuacán Caves. In *The Prehistory of the Tehuacán Valley,* vol. 1, *Environment and Subsistence,* edited by Douglas Byers, pp. 212–19. University of Texas Press, Austin.

1967 Maize of Caldwell Village (A Utah Fremont Site). In *Caldwell Village,* by J. Richard

Ambler, pp. 113–18. University of Utah Anthropological Papers No. 84, Salt Lake City.
- 1969 A New Species of *Cucurbita* from Ecuador. *Annals of the Missouri Botanical Garden* (St. Louis) 55(30): 392–96.
- 1969 *Prehistoric Cucurbits from the Valley of Oaxaca, Mexico*. Abstracts of the Eleventh International Botanical Congress, Seattle, p. 236.

Whitaker, Thomas W., and Hugh C. Cutler
- 1965 Cucurbits and Cultures in the Americas. *Economic Botany* 19(4): 344–49.
- 1967 Pottery and *Cucurbita* species. *American Antiquity* 32(2): 225–26.
- 1968 *Prehistoric Distribution of Cucurbita L. in the Americas: Unsolved Problems*. Congresso Internacional de Americas XXXX, Buenos Aires.
- 1971 Prehistoric Cucurbits from the Valley of Oaxaca. *Economic Botany* 25:123–27.
- 1986 Cucurbits from Preceramic Levels at Guila Naquitz. In *Guila Naquitz*, by K. Flannery, pp. 275–79. Academic Press, New York.

Whitaker, T. W., H. C. Cutler, and R. S. MacNeish
- 1957 Cucurbit Materials from Three Caves near Ocampo, Tamaulipas. *American Antiquity* 22(4): 352–58.

Works Cited

Anderson, Edgar
    1944   Maiz Reventador. *Annals of the Missouri Botanical Garden* (St. Louis) 31:301–15.
    1946   Maize in Mexico: A Preliminary Survey. *Annals of the Missouri Botanical Garden* (St. Louis) 33:147–247.
Anderson, Edgar, and F. D. Blanchard
    1952   Prehistoric Maize from Canyon del Muerto. *American Journal of Botany* 29:832–35.
Anderson, Edgar, and William L. Brown
    1952   The History of the Common Maize Varieties of the United States Corn Belt. *Agricultural History* 26:2–8.
Anderson, Edgar, and Hugh C. Cutler
    1942   Races of *Zea mays:* I. Their Recognition and Classification. *Annals of the Missouri Botanical Garden* (St. Louis) 29:69–89.
Armour, David A.
    1978   *Attack at Michilimackinac, Alexander Henry's Travels and Adventures in Canada and the Indian Territories between the Years 1760 and 1764.* Mackinac Island State Park Commission, Mackinac Island, Michigan.
Armour, David A., and Keith R. Widder
    1978   *At the Crossroads, Michilimackinac during the American Revolution.* Mackinac Island State Park Commission, Mackinac Island, Michigan.
Beadle, George W.
    1972   *The Mystery of Maize.* Field Museum of Natural History Bulletin, vol. 43, no. 10, Chicago, Illinois.
Benson, Adolph B. (editor)
    1966   *Travels in North America by Peter Kalm*, vol. 1. Dover Publications, New York.
Binford, Lewis R.
    1967   Smudge Pits and Hide Smoking: The Use of Analogy in Archaeological Reasoning. *American Antiquity* 32:1–12.
    1972   Archaeological Reasoning and Smudge Pits—Revisited. In *An Archaeological Perspective*, edited by L. R. Binford, pp. 52–58. Seminar Press, New York.
Blake, Leonard W.
    1981   Early Acceptance of Watermelons by Indians of the United States. *Journal of Ethnobiology* 1(2): 193–99.
    1986   Corn and Other Plants from Prehistory into History in Eastern United States. In *The Protohistoric Period in the Mid-South: 1500–1700: Proceedings of the 1983 Mid-South Archaeological Conference*, edited by David H. Dye and Ronald C.

　　　　Brister, pp. 3–13. Archaeological Report No. 18, Mississippi Department of Archives and History, Jackson.
Blake, Leonard W., and Hugh C. Cutler
　　1975　Food Plant Remains from the Zimmerman Site. Appendix 2 in *The Zimmerman Site*, by Margaret Kimball Brown, pp. 92–94. Illinois State Museum Reports of Investigations No. 32, Springfield.
　　1979　Plant Remains from the Upper Nodena Site (3Ms4). *Bulletin of the Arkansas Archaeological Society* 20:53–58.
　　1982　Plant Remains from the King Hill Site (23Bn1) and Comparisons with Those from the Utz Site (23Sa2). *Missouri Archaeologist* (Columbia) 43:101–2.
Bohrer, Vorsila L.
　　1960　Zuni Agriculture. *El Palacio, Journal of the Museum of New Mexico* (Albuquerque) 62(6): 181–202.
Bourne, Edward Gaylord (editor)
　　1904　*Narratives of the Career of Hernando DeSoto*, translated by Buckingham Smith. A. S. Barnes, New York.
Bradfield, Maitland
　　1971　*The Changing Pattern of Hopi Agriculture*. The Royal Anthropological Institute of Great Britain and Ireland, London.
Bray, Robert T.
　　1971　*The Stone Lined Privy at Hanley House*. Manuscript, University of Missouri, Columbia (Mimeographed).
Brooks, Richard H., L. Kaplan, Hugh C. Culter, and T. W. Whitaker
　　1962　Plant Remains from Cave on the Rio Zape, Durango, Mexico. *American Antiquity* 27:356–69.
Brown, Margaret Kimball
　　1975　*The Zimmerman Site*. Illinois State Museum Reports of Investigations No. 32, Springfield.
Brown, William L., and Edgar Anderson
　　1947　The Northern Flint Corns. *Annals of the Missouri Botanical Garden* (St. Louis) 34:1–29.
　　1948　The Southern Dent Corns. *Annals of the Missouri Botanical Garden* (St. Louis) 35:255–74.
Brown, William L., E. G. Anderson, and Roy Tuchawena
　　1952　Observations on Three Varieties of Hopi Maize. *American Journal of Botany* (Baltimore) 39(8): 597–60.
Carter, George F.
　　1945　*Plant Geography and Culture History in the American Southwest*. Wenner-Gren Foundation for Anthropological Research, Viking Fund Publications in Anthropology No. 5, pp. 1–140, New York.
Carter, George F., and Edgar Anderson
　　1945　A Preliminary Survey of Maize in the Southwestern United States. *Annals of the Missouri Botanical Garden* (St. Louis) 32:297–323.
Castetter, Edward F., and Willis H. Bell
　　1942　*Pima and Papago Indian Agriculture*. Inter-Americana Studies 1, University of New Mexico Press, Albuquerque.
　　1951　*Yuman Indian Agriculture*. University of New Mexico Press, Albuquerque.

Chapman, Carl H.
 1974   *Osage Indians III, The Origin of the Osage Indian Tribe*. Garland Publishers, New York.

Clark, J. Allen
 1956   Collection, Preservation, and Utilization of Indigenous Strains of Maize. *Economic Botany* 10(2): 194–200.

Conard, Nicholas, D. L. Asch, N. B. Asch, D. Elmore, H. Gove, M. Rubin, J. A. Brown, M. D. Wiant, K. B. Farnsworth, and T. G. Cook
 1984   Accelerator Radiocarbon Dating of Evidence for Prehistoric Horticulture in Illinois. *Nature* 308:443–46.

Connor, Jeannette Thurber (editor and translator)
 1925   *Colonial Records of Spanish Florida*. Vol. 1, *Letters and Reports of Governors and Secular Persons, 1570–1577*. Deland, Florida.

Coon, Carlton S.
 1962   *The Origin of Races*. Alfred A. Knopf, New York.

Cutler, Hugh C.
 n.d.   *Corn from the Great Osage Site*. Manuscript on file at the Missouri Botanical Garden, St. Louis.
 1952   A Preliminary Survey of Plant Remains of Tularosa Cave. In *Mogollon Cultural Continuity and Change: Stratigraphic Analysis of Tularosa and Cordova Caves*, edited by P. S. Martin, J. B. Rinaldo, E. Bluhm, H. C. Cutler, and R. Grange, Jr., pp. 461–79. *Fieldiana: Anthropology* (Chicago) 40:1–529.
 1960   Cultivated Plant Remains from Waterfall Cave, Chihuahua. *American Antiquity* 26:277–79.
 1965a Prehistoric Plant Remains from the Gila Bend Area. In *Salvage Archaeology in the Painted Rocks Reservoir, Southern Arizona*, by William W. Wasley and Alfred E. Johnson. University of Arizona Anthropological Papers 9, pp. 108–9, Tucson.
 1965b Vegetal Remains: Corn and Cucurbits. In *A Survey and Excavation of Caves in Hidalgo County, New Mexico*, by Marjorie F. Lambert and J. Richard Ambler, pp. 90–94. School of American Research Monograph No. 25, Santa Fe.
 1966   *Corn, Cucurbits and Cotton from Glen Canyon*. Anthropological Papers No. 80, pp. 1–62, Department of Anthropology, University of Utah, Salt Lake City.
 1968   Origins of Agriculture in the Americas. *Latin American Research Review* 3(4): 3–21.
 1980   Corn from 38Bu162A. Appendix A in *The Discovery of Santa Elena*, by Stanley South. Research Manuscript Series 165, Institute of Archaeology and Anthropology, University of South Carolina, Columbia.

Cutler, Hugh C., and Leonard W. Blake
 1971   Travels of Corn and Squash. In *Man Across the Sea*, edited by C. L. Riley, J. C. Kelley, C. W. Pennington, and R. L. Rands, pp. 366–75. University of Texas Press, Austin.
 1976   *Plants from Archaeological Sites East of the Rockies*. American Archaeology Reports No. 1, American Archaeology Division, University of Missouri, Columbia (Microfiche). (Also in this volume.)

Cutler, Hugh C., and Marion C. Cutler
 1948   Studies on the Structure of the Maize Plant. *Annals of the Missouri Botanical Garden* (St. Louis) 35:301–16.

Cutler, Hugh C., and Winton Meyer
 1965   Corn and Cucurbits from Wetherill Mesa Sites. *American Antiquity* 31(2): 136–52.
Cutler, Hugh C., and Thomas W. Whitaker
 1961   History and Distribution of the Cultivated Cucurbits in the Americas. *American Antiquity* 26(4): 469–85.
 1967   Cucurbits from the Tehuacán Caves. In *The Prehistory of Tehuacán Valley*, vol. 1, edited by Douglas S. Beyers, pp. 212–19. University of Texas Press, Austin.
de Wet, J. M. J., and J. R. Harlan
 1971   Origin of Maize: The Tripartite Hypothesis. *Eupytica* 21:271–79.
Dick, Herbert W.
 1965   *Bat Cave*. School of American Research Monograph No. 27, pp. 1–114, Santa Fe.
Doebley, John F., Major M. Goodman, and Charles W. Stuber
 1986   Exceptional Genetic Divergence of Northern Flint Corn. *American Journal of Botany* 73(1): 64–69.
Driver, Harold E., and William C. Massey
 1957   *Comparative Studies of North American Indians*. Transactions of the American Philosophical Society, new series 47, pt. 2, Philadelphia.
Ellsworth, Henry Leavitt
 1937   *Washington Irving on the Prairie*, discovered and edited by Stanley T. Williams and Barbara D. Simison. American Book, New York.
Emerson, R. A., and Harold H. Smith
 1950   *Inheritance of Number of Kernel Rows in Maize*. Cornell University Agricultural Experiment Station, Memoir 296, Ithaca, New York.
Ewan, Joseph, and Nesta Ewan
 1970   *John Banister and His Natural History of Virginia, 1678–1692*. University of Illinois Press, Urbana.
Fernald, M. L. (editor)
 1908   *Gray's New Manual of Botany*. 7th ed. American Book, New York.
Fernald, Merritt Lyndon, and Alfred Charles Kinsey
 1958   *Edible Wild Plants of Eastern North America*. Harper & Row, New York.
Forde, Cyril Daryll
 1963   *Habitat, Economy, and Society: A Geographical Introduction to Ethnology*. E. P. Dutton, New York.
Fritz, Gayle J.
 1994   Precolumbian *Cucurbita argyrosperma* (Cucurbitaceae) in the Eastern Woodlands of North America. *Economic Botany* 48(3): 280–92.
Galinat, W. C.
 1965   The Evolution of Corn and Culture in North America. *Economic Botany* 19(4): 350–57.
Galinat, Walton C., and James H. Gunnerson
 1961   *Spread of Eight-Rowed Maize from the Prehistoric Southwest*. Botanical Museum (Harvard) Leaflets 20(5): 117–60.
Gates, Charles M. (editor)
 1933   *Five Fur Traders of the Northwest, Being the Narrative of Peter Pond and the Diaries of John Macdonell, Archibald N. McLeod, Huqh Faries and Thomas Connor*. Minnesota Society of Colonial Dames of America, University of Minnesota Press, Minneapolis.

Gerin-Lajoie, Marie (translator and editor)
    1976   *Fort Michilimackinac in 1749, Lotbiniere's Plan and Description.* Mackinac History, vol. 2, leaflet no. 5, Mackinac Island State Park Commission, Mackinac Island, Michigan.

Gilmore, Melvin Randolph
    1919   *Uses of Plants by the Indians of the Missouri River Region.* Thirty-third Annual Report of the Bureau of American Ethnology, pp. 45–154, Government Printing Office, Washington, D.C.

Glover, Richard (editor)
    1962   *David Thompson's Narrative, 1784–1812.* The Champlain Society, Toronto.

Heckwelder, John
    1971   *History, Manners and Customs of the Indian Nations Who Once Inhabited Pennsylvania and the Neighboring States* [1876]. Reprinted from a copy in the State Historical Society of Wisconsin Library, Arno Press and New York Times, New York.

Hedrick, U. P., W. T. Tapley, G. P. Van Eseldine, and W. D. Enzie
    1931   *Beans of New York.* Vol. 1, pt. 2 of *Vegetables of New York.* Albany, New York.

Heldman, Donald P.
    1983   *Archaeological Investigations at French Farm Lake in North Michigan, 1981–1982: A British Colonial Farm Site.* Archaeological Completion Report Series No. 6, Mackinac Island State Park Commission, Mackinac Island, Michigan.

Henry, Alexander
    1901   *Travels and Adventures in Canada and the Indian Territories between the Years 1760 and 1776,* edited by James Bain. New edition. Little Brown, Boston.

Innis, Harold A.
    1962   *The Fur Trade in Canada.* Yale University Press, New Haven.

Jackson, Donald, ed.
    1966   *The Journals of Zebulon Montgomery Pike with Letters and Related Documents.* 2 vols. Norman, Oklahoma.

Kaplan, Lawrence
    1956   The Cultivated Beans of the Prehistoric Southwest. *Annals of the Missouri Botanical Garden* (St. Louis) 43:189–251.

Katz, S. H., M. L. Hediger, and L. A. Valleroy
    1974   Traditional Maize Processing Techniques in the New World. *Science* 184:765–73.

Kinietz, W. Vernon
    1965   *The Indians of the Western Great Lakes, 1615–1760.* University of Michigan Press, Ann Arbor.

Leonard, Warren H., and John R. Martin
    1963   *Cereal Crops.* MacMillan Company, New York.

Long, Austin, B. F. Benz, D. J. Donahue, A. J. T. Jull, and L. J. Toolin
    1989   First Direct AMS Dates on Early Maize from Tehuacán, Mexico. *Radiocarbon* 31(3): 1035–40.

Longley, A. E.
    1938   Chromosomes of Maize from North American Indians. *Journal of Agricultural Research* 56:177–95.

Lyon, Eugene
    1976   *The Enterprise of Florida.* University Presses of Florida, Gainesville.

McDermott, John Francis
    1941   *A Glossary of Mississippi Valley French, 1673–1850.* Washington University Studies—New Series, Language and Literature No. 12, St. Louis, Missouri.

Mackenzie, Alexander
    1966   *Voyages from Montreal on the River St. Laurence.* Photocopy of London edition of 1801. Facsimile Series No. 52, University Microfilms, Ann Arbor, Michigan.

Mangelsdorf, Paul C.
    1974   *Corn, Its Evolution and Improvement.* Belknap Press of Harvard University Press, Cambridge, Massachusetts.

Mangelsdorf, P. C., H. W. Dick, and J. Comara-Hernandez
    1967   *Bat Cave Revisited.* Botanical Museum (Harvard) Leaflets 22:1–31.

Mangelsdorf, Paul C., and R. H. Lister
    1956   *Archaeological Evidence on the Evolution of Maize in Northeastern Mexico.* Botanical Museum (Harvard) Leaflets 14:151–78.

Mangelsdorf, P. C., R. S. MacNeish, and W. C. Galinat
    1967a   *Prehistoric Maize, Teosinte, and Tripsacum from Taumalipas, Mexico.* Botanical Museum (Harvard) Leaflets 22:33–62.
    1967b   Prehistoric Wild and Cultivated Maize. In *The Prehistory of Tehuacán Valley*, vol. 1, edited by Douglas S. Beyers, pp. 178–200. University of Texas Press, Austin.

Martin, Paul S., John B. Rinaldo, Elaine Bluhm, Hugh C. Cutler, and Roger Grange, Jr.
    1952   Mogollon Cultural Continuity and Change. *Fieldiana: Anthropology* (Chicago) 40:1–528.

Miller, Judith
    1973   Genetic Erosion: Crop Plants Threatened by Government Neglect. *Science* 182:1231–33.

Munson, Patrick J.
    1969   Comments on Binford's *Smudge Pits and Hide Smoking: The Use of Analogy in Archaeological Reasoning. American Antiquity* 34:83–85.

Neuffer, M. G., Loring Jones, and Marcus S. Zuber
    1968   *The Mutants of Maize: A Pictorial Survey in Color of the Useable Mutant Genes in Maize with Gene Symbols and Linkage Map Positions, Arranged by Chromosome Location.* Crop Science Society of America, Madison, Wisconsin.

Nickerson, Norton H.
    1953   Variation in Cob Morphology among Certain Archaeological and Ethnological Races of Maize. *Annals of the Missouri Botanical Garden* (St. Louis) 40:79–111.

Parker, Arthur C.
    1910   *Iroquois Uses of Maize and Other Food Plants.* New York State Museum, Museum Bulletin 144, Education Department Bulletin No. 482, University of the State of New York, Albany.

Pennington, Campbell W.
    1969   *The Tepehuan of Chihuahua: Their Material Culture.* University of Utah Press, Salt Lake City.

Pond, Peter
    1900   *Wisconsin Historical Collections*, vol. 18, p. 328. Wisconsin Historical Society, Madison.

Quaife, Milo M. (editor)
   1928   *The John Askin Papers, Vol. 1: 1747–1795.* Burton Historical Records, Detroit Public Library, Detroit, Michigan.
Rick, Charles M., and Edgar Anderson
   1949   On Some Uses of Maize in the Sierra Ancash. *Annals of the Missouri Botanical Garden* (St. Louis) 336:405–12.
Robbins, Wilfred W., John P. Harrington, and Barbara Freire-Marreco
   1916   *Ethnobotany of the Tewa Indians.* Bureau of American Ethnology Bulletin 55, Smithsonian Institution, Washington, D.C.
Sauer, Carl Ortwin
   1941   The Personality of Mexico. *Geographical Review* 32:353–64.
Sheldon, Elizabeth Shepard
   1978   *Introduction of the Peach* (Prunus persica) *to the Southeastern United States.* Paper presented at the 19th Annual Meeting of the Society for Economic Botany, St. Louis.
Stevens, Sylvester K., Donald Kent, and Emma Edith Woods (editors)
   1941   *Travels in New France by J.C.B.* The Pennsylvania Historical Commission, Department of Public Instruction, Commonwealth of Pennsylvania, Harrisburg.
Stevenson, Matilda Cox
   1915   *Ethnobotany of the Zuni Indians.* Thirtieth Annual Report of the Bureau of American Ethnology, 1908–1909, Government Printing Office, Washington, D.C.
Stoddard, Amos
   1812   *Sketches, Historical and Descriptive of Louisiana.* Carey, Philadelphia.
Stothers, David M.
   1976   The Princess Point Complex. In *The Late Prehistory of the Lake Erie Drainage Basin: A 1972 Symposium Revised,* edited by David S. Brose, pp. 137–61. Cleveland Museum of Natural History, Cleveland, Ohio.
Struever, Stuart
   1968   Flotation Techniques for Recovery of Small-Scale Archaeological Remains. *American Antiquity* 33:353–62.
Sturtevant, E. L.
   1899   *Varieties of Corn.* U.S.D.A. Official Experimental Station Bulletin 57, Government Printing Office, Washington, D.C.
Swanton, John R.
   1911   *Indian Tribes of the Lower Mississippi Valley and Adjacent Coast of the Gulf of Mexico.* Bureau of American Ethnology, Bulletin 43, Smithsonian Institution, Washington, D.C.
   1946   *The Indians of the Southeastern United States.* Bureau of American Ethnology Bulletin 137, Smithsonian Institution, Washington, D.C.
Temple, Wayne C.
   1958   *Indian Villages of the Illinois Country, Historic Tribes.* Scientific Papers, vol. 2, pt. 2, Illinois State Museum, Springfield.
Thwaites, Ruben Gold (editor)
   1897–1900   *Jesuit Relations and Allied Documents,* vol. 59. Burrows Brothers, Cleveland, Ohio.
Tooker, Elizabeth
   1964   *An Ethnography of the Huron Indians, 1615–1649.* Bureau of American Ethnology Bulletin 190, Smithsonian Institution, Washington, D.C.

Upham, Steadman, Richard R. MacNeish, Walton C. Galinat, and Christopher M. Stevenson
    1987   Evidence Concerning the Origin of Maiz de Ocho. *American Anthropologist* 89:410–19.

U.S. Department of Agriculture
    1941   *Climate and Man.* Yearbook of Agriculture, U.S. Government Printing Office, Washington, D.C.

Varner, John, and Jeanette Varner (translators)
    1951   *The Florida of the Inca,* by Garcilaso de la Vega. University of Texas Press, Austin.

Waugh, F. W.
    1916   *Iroquois Foods and Food Preparation.* Canada Department of Mines Geological Survey, Memoir 86, No. 12, Anthropological Series, Government Printing Bureau, Ottowa.

Wellhausen, E. J., L. M. Roberts, Efram Hernandez Xolocotzi, and Paul C. Mangelsdorf
    1952   *Races of Maize in Mexico: Their Origin, Characteristics, and Distribution.* Bussey Institute, Harvard University, Cambridge, Massachusetts.

Weltfish, Gene
    1971   *The Lost Universe.* Ballantine Books, New York.

Whitaker, Thomas W., and Hugh C. Cutler
    1971   Prehistoric Cucurbits from the Valley of Oaxaca. *Economic Botany* 25:123–27.

Whitaker, Thomas W., and Glen N. Davis
    1962   *Cucurbits, Botany, Cultivation and Utilization.* World Crop Books, Interscience Publishers, New York.

Whiting, Alfred F.
    1939   *Ethnobotany of the Hopi.* Museum of Northern Arizona Bulletin No. 15, pp. 1–120, Flagstaff.

Wilkes, H. G.
    1967   *Teosinte: The Closest Relative of Maize.* Bussey Institute, Harvard University, Cambridge, Massachusetts.
    1971   Maize and Its Wild Relatives. *Science* 177:1071–77.

Will, George F., and George E. Hyde
    1968   *Corn among the Indians of the Upper Missouri* [1917]. Reprint, University of Nebraska Press, Lincoln.

Wills, W. H.
    1995   Archaic Foraging in the Beginning of Food Production in the American Southwest. In *Last Hunters, First Farmers,* edited by T. D. Price and A. B. Gebauer, pp. 215–42. School of American Research Press, Santa Fe.

Wilson, Gilbert Livingstone
    1977   *Agriculture of the Hidatsa Indians: An Indian Interpretation* [1917]. Reprints in Anthropology, J & L Reprint Co., Lincoln, Nebraska. Reprint of Minnesota Studies in Social Sciences No. 9, Minneapolis.

Yanovsky, Elias
    1936   *Food Plants of the North American Indians.* U.S. Department of Agriculture, Miscellaneous Publication No. 237, Washington, D.C.

Yarnell, Richard Asa
    1964   *Aboriginal Relationships between Culture and Plant Life in the Upper Great Lakes Area.* Anthropological Papers No. 23, Museum of Anthropology, University of Michigan, Ann Arbor.

Index of Latin Names for Plant Taxa

Alnus sp., 111
Amaranthus sp., 100, 103, 107, 108, 129, 144
Ambrosia alternifolia, 140
Ambrosia sp., 100, 115, 122, 126, 127
Ampelopsis sp., 118, 121
Apios americana, 98, 102, 105, 109, 112, 121, 122
Apocynum sp., 100
Arundinaria gigantean, 100, 101, 107, 109
Asimina triloba, 73, 85, 86, 87, 90, 91, 99, 103, 105, 106, 109, 119, 120, 121, 123, 126
Asteraceae, 111
Avena sativa, 2

Betula sp., 111
Brassica sp., 116

Carex sp., 103, 105, 108, 110
Carya illinoensis, 99, 102, 103, 105, 106, 107, 109, 118, 119, 120, 121, 122, 125, 142
Carya laciniosa, 109, 121, 123
Carya sp., 98, 99, 100, 102, 103, 104, 105, 106, 107, 108, 109, 110, 116, 118, 119, 120, 121, 122, 123, 124, 125, 128, 130, 134, 135, 136, 142, 144, 145, 146
Castenea dentata, 107, 135, 136
Celtis sp., 73, 84, 87, 91, 98, 99, 100, 103, 110, 111, 113, 120, 121, 126, 127, 135, 137, 139, 141
Chenopodium sp., 100, 103, 105, 107, 108, 110, 119, 125, 135, 138, 140
Cirsium sp., 124
Citrullus lanatus, 52, 53, 82–84, 87, 88, 89–90, 91, 103, 118, 120, 126, 147; varieties of, 82
Cleome serrulata, 140
Compositae. *See* Asteraceae
Corylus americana, 85, 90, 91, 98, 103, 108, 109, 110, 120, 121, 123, 125, 126, 127, 132, 136, 145, 147
Crataegus sp., 84, 85, 87, 90, 91, 98, 103, 108, 120, 122, 135, 136, 147
Cucurbita argyrosperma, 35, 36n, 92n, 97, 129, 139
Cucurbita foetidissima, 133, 134
Cucurbita maxima, 35, 97, 129, 131
Cucurbita mixta. See Cucurbita argyrosperma
Cucurbita moschata, 35, 97, 118, 131
Cucurbita pepo, 35, 48, 50, 52, 80–82, 92, 97, 100, 101, 103, 107, 110, 113, 114, 116, 120, 123, 125, 127, 128, 129, 131, 135, 137, 138, 139, 140, 141, 145, 147
Cucurbita spp., 2, 105
Cucurbita texana, 50
Cyperaceae, 108, 136

Desmodium sp., 87, 88, 99, 108
Diospyros virginiana, 98, 99, 100, 102, 104, 105, 106, 107, 109, 113, 116, 118, 119, 120, 121, 122, 123, 125, 126, 135, 136, 142, 144

Eleocharis sp., 105
Equisetum sp., 105
Euchlaena mexicana, 2, 96
Euonymus sp., 120

Fabaceae, 128, 143, 146

Gleditsia triacanthus, 100, 101, 107, 108, 109, 118, 124, 144
Glycine max, 87, 88
Gymnocladus dioca, 121

Helianthus annuus, 100, 103, 104, 106, 110,

111, 113, 118, 120, 121, 122, 123, 125, 128, 132, 133, 138, 141, 142

Ilex sp., 136
Ipomoea sp., 126
Iva sp., 100, 110, 113, 121, 122, 123, 125, 126, 128, 133, 119

Juglans cinerea, 120, 134, 135, 136, 142, 144, 146, 147
Juglans nigra, 85, 87, 90, 91, 98, 99, 100, 102, 103, 104, 106, 107, 109, 110, 112, 119, 120, 121, 122, 123, 125, 127, 128, 135, 144
Juniperus sp., 106, 107, 110, 126, 138, 139, 143, 147

Lagenaria siceraria, 50, 97, 101, 107, 114, 115, 116, 120, 124, 128, 129, 134, 135
Leguminosae. *See* Fabaceae
Lens culinaris, 2
Lithospermum caroliniense, 141
Lithospermum sp., 99, 127

Malus sp., 104, 120, 123, 125

Nelumbo lutea, 85, 87, 90, 91, 103, 120
Nicotiana rustica, 131
Nymphaea tuberosa, 109
Nyssa sylvatica, 121, 136

Opuntia engelmannii, 143
Opuntia sp., 134, 143
Oryza sp., 56, 58

Parthenocissus quinquefolia, 122
Paspalum sp., 99
Passiflora incarnata, 122, 142
Passiflora sp., 121
Phalaris sp., 100
Phaseolus acutifolius, 23
Phaseolus polystachyus, 100
Phaseolus spp., 2
Phaseolus vulgaris, 36, 38, 48, 49, 78, 79, 80, 112
Phragmites sp., 105
Physalis sp., 144
Phytolacca americana, 86, 87
Pinus sp., 133, 144

Pisum sativum, 112
Poaceae, 105, 106, 107, 108, 109, 127, 137, 140, 142
Polygonum pennsylvanicum, 115
Populus deltoides, 127
Prosopis sp., 129
Prunus americana, 84, 110, 112, 131, 140, 141, 145, 147
Prunus angustifolia, 104
Prunus avium, 52
Prunus hortulena, 104, 109, 112, 140
Prunus nigra, 147
Prunus pennsylvanicum, 117
Prunus persica, 36, 52, 53, 98, 120, 129, 135, 137
Prunus serotina, 90, 99, 103, 105, 107, 109, 110, 115, 121, 127, 135, 137, 144
Prunus sp., 73, 84, 87, 90, 91, 102, 103, 104, 106, 107, 109, 110, 112, 116, 118, 120, 121, 122, 126, 130, 131, 132, 135, 136, 137, 138, 139, 144, 146, 147
Prunus virginiana, 85, 87, 91, 140, 141
Psoralea esculenta, 113, 128
Pyrus sp. *See Malus* sp.

Quercus sp., 37, 98, 100, 101, 102, 103, 104, 105, 107, 108, 109, 115, 116, 118, 119, 120, 121, 122, 123, 124, 126, 130, 135, 136, 142, 144, 146, 147

Rhus sp., 144
Robinia sp., 109
Rosa setigera, 141
Rubus sp., 85, 87, 103, 104, 120

Scirpus sp., 103, 105, 108, 110, 118, 121, 145
Scirpus validus, 105
Secale cereale, 2
Shepardia canadensis, 120, 126
Smilax sp., 116
Solidago sp., 111, 122
Strophostyles helvola, 106
Strophostyles sp., 118, 119, 121
Strophostyles umbellata, 118

Tilia sp., 121, 135, 141, 145
Tripsacum dactyloides, 22
Tripsacum spp., 2, 3, 22, 96

Triticum aestivum, 2
Typha sp., 105

Viburnum prunifolium, 134
Viburnum sp., 102, 103, 104, 107, 108, 115, 116, 121, 126
Vitis sp., 85, 87, 90, 91, 99, 100, 103, 104, 107, 110, 115, 116, 117, 118, 119, 120, 121, 125, 126, 135, 138, 139

Xanthium sp., 100

Zea mays. See corn
Zea mexicana, 2

Index of Corn Races and Varieties

Baby Rice, 11
Basketmaker, 10, 114
Blando de Sonora, 8, 9, 10

Chapalote, 8, 9, 22, 134
Conical Dent, 8
Cónico, 11, 12
Cónico Norteño, 23
Corn Belt Dent, 8, 11, 13, 96
Cristalina, 8, 11, 12, 13

dent corn, 4, 10, 11, 12, 13, 23, 24, 25, 28, 70, 88, 96, 97, 126

Eastern Eight Row, xii, 4, 8, 9, 11, 12, 15, 40, 42, 46, 48, 52, 54, 67, 74–75, 77, 96, 120
Eastern flint, 98

flint corn, 6, 10, 12, 22, 24, 25, 28, 40, 54, 67, 75, 122, 125, 139, 143
flour corn, 5, 6, 22, 23, 24, 25, 28, 40, 54, 67, 75, 139
Freemont Dent, 11

Golden Bantam Sweet Corn, 11, 13

Harinoso de Ocho, 10, 22, 28

Japanese Hulless, 9, 11

Kokoma Corn, 22

Maiz de Ocho, 10, 54
Mexican Dent, 106, 143

Mexican Pyramidal, 11, 23
Midwest Twelve Row, 8, 11, 96

North American Pop, 8, 9, 48, 96
Northern Flint, 11, 40, 46, 59, 60, 61, 62–63, 67, 74, 96, 101, 103, 106, 123, 124, 127, 134, 135, 139, 144, 145

Olotillo, 12
Onaveño, 8, 9, 10, 22

Pepitilla, 11, 12, 23
Pima-Papago, 2, 8, 9–10, 11, 12, 13, 22, 23, 96
pop corn, 8, 11, 13–14, 23, 103, 105, 110, 120, 121, 133, 139, 141, 145
Pre-Chapalote, 8
Primitive Pop, 3
Pueblo, 8, 10, 11, 12, 13
Pyramidal Dent, 3

Reventador, 8, 9, 22, 96

Small Cob, 8, 9
South American Golden Pop, 9
Southern Dent, 12, 48, 52, 70, 71, 75, 120, 143
Strawberry Pop, 11

Toluca Pop, 8, 9, 11, 23
Tropical Flint, 2, 9, 96, 106, 132
Tuxpeño, 12

Walpi Blue, 6
Western Eight Row, 8, 10

Zapelote Chico, 11, 12, 23

General Index

Acoma Pueblo, 9, 12, 13, 14, 24
acorns, 37, 98, 100, 101, 102, 103, 104, 105, 107, 108, 109, 115, 116, 118, 119, 120, 121, 122, 123, 124, 126, 130, 135, 136, 142, 144, 146, 147
Adena. *See* Daines II Adena Mound
Alabama, 52, 70, 97, 98
alcohol, 17
alder, 111
aleurone, 4, 5
Alhart site, 62
allowances of food for voyageurs, 55
amaranth, 100, 103, 107, 108, 129, 144
Aminoya, 37–39
AMS dating, 1
Anasazi. *See* Ancestral Pueblo
Ancestral Pueblo, 26
Andes, 35
Antelope Cave, Arizona, 26
Antelope House, Canyon de Chelly, 12, 26
Apache people, 12
Arbre Croche. *See* L'Arbre Croche
Archaic Period, 40, 41
Arikara people, 62
Arizona, 3, 6, 8, 9, 10, 12, 16, 22, 23, 34, 35, 36
Arkansas, 2, 9, 77, 99–100
Arkansas River, 70
ash, in food preparation, 55, 56
Askin, John, 57, 58, 61
Atlantic coast, xii, 9

Bamble site, 62
Banister, John, 42, 97
Bareis, Charles, 80
Basketmaker Period, 10, 26, 114
basswood, 121, 135, 141, 145
Bat Cave, 2, 3, 8, 22, 96
bead grass, 99
Beadle, George W., 2

bean, 2, 23, 36, 38, 48, 49, 64, 73, 78, 82, 84, 87, 89–90, 91, 100, 101, 102, 103, 104, 110, 112, 113, 114, 118, 120, 122, 123, 127, 128, 129, 130, 131, 132, 133, 135, 136, 137, 138, 139, 140, 142, 143, 145, 146; wild, 38, 100, 106
beeweed, 140
Benchmark Cave, 27
Betatakin, 12
Big Osage site, xv. *See also* Carrington site
birch, 111
blackberry, 86, 87, 103
black gum, 121, 136
black haw, 102, 103, 104, 107, 108, 115, 116, 121, 126, 134
Blake, Leonard, xi, xiii–xiv
bluestem grass, 140
Bois Blanc Island, 60
Bolivia, 13
bottle gourd, 3, 35, 50, 97, 101, 107, 114, 115, 116, 120, 124, 128, 129, 134, 135
Bray, Robert T., 86
British, 52, 75, 84, 97
Brown, James, 59
Brown, Margaret K., 66, 68, 79
Brown site, 47, 48, 49, 51, 52, 53
buffalo berry, 120, 126
buffalo gourd, 133, 134
butternut, 120, 134, 135, 136, 142, 144, 146, 147

cabbage, 116
cactus, 134, 143
Cahokia Mound, 2, 36n, 66, 80, 81
Caldwell Village, 27
California, 52
Canada, 6, 11, 46, 52, 54, 60, 62, 84, 96
canary grass, 100
cane, 100, 101, 107, 109
Canyon de Chelly, 10, 12, 36

Carbon 14 dating, 1, 3, 60
carbonization, 24, 48, 78
carbonized plants, 24, 46, 52, 60, 73, 77, 84
Carmona, Alonso de. *See* de Carmona, Alonso
Carrington site, xv, 47, 48, 49, 51, 52
Carter Ranch Pueblo, 26
Cartier, Jacques, 59
Casper Cliff Ruin, 26
catbrier, 116
cattail, 105
Champlain, Samuel D., 16, 60
Chapman, Carl H., 46, 52
Cherokee people, 11
cherry, domestic sweet, 52
cherry, ground, 144
cherry, wild, 90, 99, 103, 105, 107, 109, 110, 115, 121, 127, 135, 137
chestnut, 107, 135, 136
Chippewa people, 58
chokecherry, 85, 87, 91, 118, 140, 141
Coal Pit site, xv, 47, 48, 49, 51, 52, 53, 63, 75, 76, 77, 83
cocklebur, 100
Cocopa, 10
coffee tree, 121
Colorado, 20, 23
Colorado River, 6, 12
Committee on Preservation of Indigenous Strains of Maize, 22
Connecticut, 85
corn, xii, 1–18, 37–39, 54–58, 87, 89–90, 91, 98–147; anatomy of, 3–4, 20; biodiversity in, 18; carbonized, 6, 14; ceremonial use of, 4, 5, 20; classification of, 7; cobs of, 73, 74; color of, 5; cultivation, 16, 57, 59, 60, 88, 91; diseases, 13; drying, 54, 55, 57; ear, 20; earliest appearance in archaeological record, 2–3; evolution over time, 6–7, 19, 20, 23; exchange, 14, 25, 72; genetics of, 5; germ plasm, 13, 18; harvest, 16; kernels of, 4, 6, 7, 11, 13, 14, 15, 23, 42, 45, 95; measurement of, 15, 20, 22, 45, 47, 64–65, 67, 74; nutritional value of, 55; planting, 57, 59, 88, 91; preparation, 16–17, 54–57; pollination of, 4, 14–15; races of, 7–14, 95; selection of seeds, 4–5, 6, 14, 16, 19–20, 22, 25; shrinkage during charring, 65; 95; storage, 16; variation in, 6; yields of, 38. *See also* Index of Corn Races and Varieties
Cornell, John, 57
corn leaf blight, 13
cottonwood, 127
crab apple, 104, 120, 123
Crawford Farm site, 63, 67, 68–69, 75, 76, 77, 85
Creek, Upper, 68, 70
Cross Village. *See* L'Arbre Croche
cupules, xii, 15, 20, 46, 74, 95, 96; width of, 20, 21, 25, 29–34, 40–45, 47, 62–63, 64–65, 67, 68, 74, 75, 76, 95
Cutler, Hugh C., xi, xii–xiii, 1, 2, 46, 93

Daines II Adena Mound, 1, 9, 94
de Carmona, Alonso, 38
Decker, Deena, 50
deer skins, 48, 77
de la Vega, Garcilaso, 38
de Soto, Hernando, 37–39
Detroit, 58, 61
Dick, Herbert, 12, 19, 24
Double Ditch site, 62
Durango, Mexico, 3, 11

Eastern North America, 4, 11, 48, 52, 70, 78, 84, 93, 97
Eastern Woodland peoples, 9
Echo Cave, 27
Ellsworth, Henry, 55
Elvas, the Gentleman of, 38
endosperm, 5, 13
English. *See* British
environmental factors influencing limits of crop plants, 19, 23, 59, 75
Europeans, 11, 12, 24, 25, 28, 35, 36, 37–39, 48, 52, 54, 57, 58, 59, 60, 61, 64, 68, 70, 74, 75, 82, 84, 97

Fairbanks, Charles, 68
false grape, 99, 119, 121
fertilization, 60
Florida, 50, 52, 82, 85, 101
flotation, xiii, 46, 50, 72, 73, 85, 94
fog, 61
Fort Berthold, 97
Fort de Chartres, 66, 68–69, 70

172   General Index

Fort Erie, 58
Fort Michilimackinac, xv, 54–58, 59–65
Fox people, 11, 66, 75, 76, 85
Fremont Culture, 12, 13, 20, 23, 27, 97
French, 52, 57, 58, 59, 60, 61, 64, 82
frost-free season, 88
fur traders, 54–58

Galisteo, New Mexico, 27
Gastonia, North Carolina, 40
Georgetown Phase, 7, 10, 22, 26
Georgia, 2, 9, 40, 42, 45, 102
Georgian Bay, 64
Gilmore, Melvin Randolph, 84, 85
Glen Canyon site, 12, 27
goldenrod, 111, 122
gourd, 20, 35, 97, 100. *See also* bottle gourd
Grande Portage, xv, 56
Gran Quivera, 12
grape, 85, 87, 90, 91, 99, 100, 103, 104, 110, 115, 116, 117, 118, 119, 120, 121, 125, 126, 135, 138, 139
grasses, 105, 106, 107, 108, 109, 127, 137, 140, 142
Great Lakes, 54–58, 59–64
Green Bay, 58
ground cherry, 144
groundnut, 98, 102, 105, 109, 112, 121, 122
Gulf Coast, 11, 97
Gumbo Point site, 47, 48, 49

hackberry, 73, 84, 87, 91, 98, 99, 100, 103, 110, 111, 113, 120, 121, 126, 127, 135, 139, 141
Hanley House, 86
haw, black, 102, 103, 104, 107, 108, 115, 116, 121, 126
hawthorn, 84, 85, 87, 90, 91, 103, 108, 120, 122, 135, 136, 147
Hayes, Alden, 12
Hayes site. *See* Coal Pit site
hazelnut, 85, 90, 91, 98, 103, 108, 109, 110, 120, 121, 123, 124, 125, 126, 127, 132, 136, 145, 147
Heiser, Charles B., 50
Henry, Alexander, 55, 58
Herold, E. B., 68
hickory, 98, 99, 100, 102, 103, 104, 105, 106, 107, 108, 109, 110, 116, 118, 119, 120, 121, 122, 123, 124, 125, 128, 130, 133, 134, 135, 136, 142, 144, 145, 146

Hidalgo County, New Mexico, 8
Hidatsa people, 57
Hinkle Park Cliff Dwelling, 26
Historic Period, 45
holly, 136
hominy, 55
Hooper Ranch Pueblo, 26
Hopewell Period sites, 1
Hopewell site, 9
Hopi Crops Survey, 22
Hopi people, 4, 5, 6, 7, 8, 9, 10, 12, 13, 14, 15, 16, 22, 23, 24, 25, 28, 34
Hulse, Frederick S., 1
Huron people, 16, 54, 59, 62, 64
hybridization, 2, 3, 5, 7, 19

Illinois, xii, 1, 2, 9, 48, 63, 72, 75, 82, 84, 93, 94, 103–108
Illinois Association for the Advancement of Archaeology, xiii, 37
Illinois confederation, 66, 67, 77, 82
Illinois State Museum, ix, 93
Indiana, 109
Inscription House, 12
insects, 25
Iowa, 110–111
Iowa people, 11
Iroquois, 6, 11, 16, 55, 59, 62, 78
Isleta Pueblo, 24, 28

Jalisco, Mexico, 35
Jasper Newman site, 1, 94
Jemez Pueblo, 4, 12, 13, 24
Jones, Volney H., 22
Josselyn, John, 84
Joutel, Henri, 77
juniper, 106, 107, 110, 126, 138, 139, 143, 147

kale, 116
Kansas, 112–13
Kansas City, 78
Kaskaskia people, 67, 78, 79, 83
Kaskaskia village, 50, 67
Kentucky, 97, 114–15
Keresan Pueblos, 24, 28
Kickapoo Historic site, xv, 47, 48, 50
Kickapoo people, 67, 75, 76, 78, 83
Kiet Siel, 27
King Hill site, 48, 50, 83, 86

General Index 173

kingnut, 109, 121, 123
Klipper, Walter, 54
knotweed, 87, 88, 107, 110, 115, 116, 136

Laguna Pueblo, 24, 28
Lake Michigan, 58, 64
Lake Superior, 56
L'Arbre Croche, 57, 58, 61
Las Madres, 27
legume, 128, 143, 146
LeJeune, Paul, 54
lentils, 2
Little Osage people, 83
Little Osage site, xv. *See also* Coal Pit site
locust, honey, 100, 101, 107, 108, 109, 118, 124, 144
Lotbiniere, Michael Chartier, 57
lotus, 85, 87, 90, 91, 103, 120
lye, 55, 56

Macdonell, John, 56
Mackenzie, Alexander, 56
Mackinac Island, 57, 59–64
Mackinaw City, Michigan, 54, 60, 61
maize. *See* corn
mallow, 107
Mandan, 59, 62, 82
Maricopa people, 10
Maricopa Reservation, 23
Marquette, Jacques, 67, 82
marshelder, 100, 110, 113, 119, 121, 122, 123, 125, 126, 128, 138
Martinez, Pat, 24
Martinez, Tom, 24
Massachusetts, 85
May, Alan, 40
maypops, 122, 142
McGraw site, 1, 94
McIvan site, 62
measurement: of bean seeds, 49–50, 78, 79, 80; of corn, 15, 20, 45, 47, 64–65, 67, 74; of squash seeds, 51, 80–82; of watermelon seeds, 82–84
Mesa Verde, 26, 27, 35
Mesita Pueblo, 24, 28
mesquite, 129
Mexico, 1, 2, 3, 6, 7, 8, 9, 11, 12, 13, 15, 16, 22, 23, 24, 35, 50, 52, 70, 75, 82, 96, 97
Michigamea people, 66, 67, 68, 70, 76, 77

Michigamea site, xv, 66–71, 76, 77
Michigan, 54–57, 59–64; University of, 22
Midwest United States, 11, 46–53, 66–71, 76, 77, 84, 96
milkweed, 100
Milwaukee, 58, 61
Minnesota, 75, 117
missions, 40, 42, 45, 97
Mississippi, 118–19
Mississippian Period, 2, 3, 40, 41, 43, 44, 45, 68, 78, 79, 80, 85
Mississippi River, 2, 9, 22, 70, 96
Mississippi Valley, xii, 10, 20, 38, 60, 96
Missouri, 16, 48, 63, 75, 85, 86, 120–26
Missouri Archaeological Society, xi, xiii, 37
Missouri Botanical Garden, xi, xii, xiii, 13, 93
Missouri Historic sites, xii, xv, 46–53
Missouri people, 46–53
Missouri River, 38, 59, 61, 82, 84, 85, 97
Moencopi, 16, 28, 34
Mogollon, 7, 22, 26
Mohave, 28, 34
Monk's Mound, 80
Montreal, 59
morning glory, 126
Morris, Donald, 12
Morse, D. F., 68
Mound City Archaeological Society, 37
Mound 51, Cahokia, 80–81. *See also* Cahokia Mound
Mummy Cave, Canyon de Chelly, 10, 26
Museum of Northern Arizona, 22

Natchez people, 82
Navajo people, 6, 12, 13, 23
Nebraska, 85, 127–28
Neutral people, 59
New England. *See* Northeastern United States
New Mexico, 3, 6, 7, 8, 10, 22, 23, 35, 36, 96, 129
New York, 6, 10, 60, 62, 64, 78, 84, 85, 130
Niagara, 58
North America, 4
North Carolina, 40–45, 52
North Dakota, 54, 57, 59, 62, 85, 131
Northeastern United States, 6, 14, 22, 42, 54, 60, 75

Northwest Company, 58
Nottaway people, 42, 45
nuts, 37, 73. *See also* butternut; hazelnut; hickory; kingnut; walnut
Nuyaka site, 68–69, 70, 97

oak. *See* acorns
oats, 2
Oaxaca, Mexico, 35, 50
O-Block Cave, 26
O'Brien, M. K., 79
Ohio, 1, 75, 94, 132
Ohio River, 70, 77
Oklahoma, 2, 9, 70, 133–34
Old World, 4, 52, 74, 82
Omaha people, 5
Oneota, 50, 83
Onondaga people, 62
Ontario, 11, 46, 62, 96
Osage people, 46–53, 55, 63, 76
Osage sites, xii, xv, 46–53. *See also* Carrington site; Coal Pit site
Ottawa people, 57–58, 61, 64

Papago people, 6, 9, 10, 12, 14, 16, 23
Parker, Arthur C., 55
passion flower, 122, 142
Pawnee, 84
pawpaw, 73, 85, 86, 87, 90, 91, 99, 103, 105, 106, 109, 119, 120, 121, 123, 126
peach, 36, 52, 53, 98, 120, 129, 135, 137
Pearsall, Deborah M., xiii
pecan, 99, 102, 103, 105, 106, 107, 109, 118, 119, 120, 121, 122, 125, 142
Peking Man, 73
Penasco Phase, 28
Pennington, Campbell, 3, 8
Pennsylvania, 10, 60, 84, 96
pericarp, 4
persimmon, 98, 99, 100, 102, 104, 105, 106, 107, 109, 113, 116, 118, 119, 120, 121, 122, 123, 125, 126, 135, 136, 142, 144
Peru, 13, 16
pests, 13, 25
Picuris, x, xv, 12, 19, 20, 21, 23–25, 27, 28, 29–33, 35
pigweed, 100, 103, 105, 107, 108, 110, 119, 125, 135, 138, 140
Pike, Zebulon Montgomery, 52

Pima people, 6, 10, 12, 23, 28
pine, 144
piñon nuts, 133
Plains of North America, 10, 22, 60, 84, 93, 96
Plains peoples, 4, 9, 14, 16, 48
Plattner site, 47, 48, 49, 50
plum, 73, 84, 87, 90, 91, 102, 103, 104, 106, 107, 109, 110, 112, 116, 120, 121, 122, 126, 130, 131, 132, 135, 136, 137, 138, 139, 144, 146, 147
pokeweed, 85, 87
Pond, Peter, 58, 60, 61, 64
population estimates, 38
Potawatomi people, 58
prairie rose, 141
prairie turnip, 113, 128
Prehistoric Period, 40, 41, 43, 45
prickly pear, 134, 143
prunes, 37
puccoon, 99, 127, 141
Puebla, Mexico, 35
Pueblo Largo, 27
Pueblo peoples, 4, 6, 10, 12, 13, 14, 15, 20, 22, 23, 24, 35, 36, 97
Pueblo Period, 12, 13, 20, 23, 26, 27, 97
Pueblo Rebellion, 12, 13, 19, 97
pumpkin, 35

radiocarbon dating. *See* AMS dating; Carbon 14 dating
ragweed, 100, 115, 122, 126, 127, 140
raspberry, 86
Red Bow Cliff Dwelling, 8
redbud, 126, 135
reeds, 100, 105, 106
Reserve Phase, 26
Rhoads Kickapoo site, xv, 47, 48, 50, 67, 72–92
rice, wild, 56, 58
Rio Grande pueblos, 4, 7, 8, 9, 12, 23, 24
Rio Grande River, 12, 97
Rio Zape, Durango, Mexico, 11
Rock River, 75
Rocky Mountains, 96
row number of corn cobs, 20, 21, 22, 25, 26–34, 40–45, 46, 47–48, 54, 62–63, 65, 67, 68, 70, 74, 75, 76, 88, 95, 96
rush, horsetail, 105

General Index 175

rushes, 103, 108, 127
rye, 2

Sagard, Gabriel, 16, 64
Salado, 26
Salts Cave, 97
San Carlos Apache Reservation, 8
San Juan River, 23
San Lorenzo Phase, 27
San Lorenzo site. *See* Picuris
Santa Fe Phase, 24, 27, 29
Santo Domingo Pueblo, 24
Sauer, Carl, 3
Sauk people, 63, 67, 68, 75, 76, 85
Sault Ste. Marie, 56, 57
Schiele Museum of Natural History, 40
sedge, 108, 136
Sheep rockshelter, 11, 63, 92
Sheldon, Elizabeth, 52
Shonto, 13
shrinkage during charring, 65, 95
Sierra de Ancash, Peru, 16
smartweed, 87, 88, 107, 110, 115, 116, 136
smudge pits, 48, 66, 73, 74, 76
Society of American Archaeology (SAA), 46
Sopher site, 62
Soto, Hernando de. *See* de Soto, Hernando
South America, 1, 11, 13, 35, 97
South Carolina, 10, 52, 70, 136
South Dakota, 62, 85, 137–41
Southeastern United States, 40–42, 70
South Park site, Wittlesey Focus, 75
Southwestern United States, 1, 8, 9, 10, 14, 16, 20, 22, 23, 26–28, 35, 75, 78, 96
soybean, 87, 88
Spanish, 24, 25, 28, 35, 36, 37–39, 40, 42, 45, 52, 70, 82, 97
spindle tree, 120
squash, 2, 3, 20, 35–36, 38, 48, 50, 52, 73, 80–82, 84, 87, 88, 89–90, 91, 97; acorn, 35, 97; banana, 35; butternut, 35; cushaw, 35, 97; Fort Berthold, 51; Hubbard, 35, 97; Mandan, 51, 82, 110, 125, 127, 131, 137, 138, 141; Mexican banana, 97; pumpkin, 97, 103, 110, 116, 125, 135, 140; Red Lodge, 110, 138; summer, 35, 82, 90, 97, 103, 116, 131; summer crookneck, 51, 97; turban, 97; white bush scallop, 35; winter 82; zucchini, 97

St. Augustine, 52, 70
St. Lawrence River Valley, 54
St. Louis, 80, 86
Ste. Genevieve, 66
Stoddard, Amos, 77
Sub-Mound 51, Cahokia, 81. *See also* Cahokia Mound
sumac, 144
sumpweed, 100, 110, 113, 119, 121, 122, 123, 125, 126, 128, 133
sunflower, 100, 103, 104, 106, 110, 111, 113, 118, 120, 121, 122, 123, 125, 128, 132, 133, 138, 141, 142
Susquehanock people, 63

Talus Ruin, 27
Tamaulipas, 3, 8
Taos Phase, 21, 27, 35
Taos Pueblo, 19, 28
Tehuacan Valley sites, 1, 2, 3, 8
Tennessee, 142
teosinte, 2, 3, 96
tepary beans, 23
Tepehuan, 17
Texas, 85–97
Thompson, David, 56
Three Circle Phases, 26
Thurman, Melburn D., 66, 68, 70
tick clover, 87, 88, 99, 108
tobacco, 131
Tonto Ruin, 26
trade, 74
Trampas Phase, 24, 28, 31, 32, 35, 36
tubers, 125
Tularosa Cave, 3, 7, 8, 10, 22, 26, 96
tupelo, 116
Turkey Cave, 26
turnip, 116

United States National Seed Laboratory, 18
Upper Creek people. *See* Creek
Upper Nodena site, 68–69, 70
Utah, 20, 23, 97
Utz site, 46, 47, 48, 49, 51, 52, 53

Vadito Phase, 27, 30, 35, 36
Vega, Garcilaso de la. *See* de la Vega, Garcilaso
Veracruz, 50

Vérendrye, Pierre, 59
Virginia, 12, 40, 42, 45, 70, 97, 144
Virginia creeper, 122
Voigt, Eric, xiii
voyageurs, xii, 54–58

Walapai, 36
walnut, 85, 87, 90, 91, 98, 99, 100, 102, 103, 104, 106, 107, 109, 110, 112, 119, 120, 121, 122, 123, 125, 127, 128, 135, 144
water lily, 109
Waterman site, xv, 66–71, 76, 77
watermelon, 52, 53, 82–84, 87, 88, 89–90, 91, 103, 118, 120, 126, 147; varieties of, 82
water screening, 74, 85, 94

West Virginia, 144
wheat, 2
Whiting, Alfred H., 22
Whittlesey Focus. *See* South Park site
Wilkinson, Lt. James B., 52
willow, 111
Wilson, Gilbert Livingstone, 14
Wisconsin, 145–46

Yampa River, 23
Yanovsky, Elias, 84, 85
Yuma people, 23

Zia Pueblo, 13
Zimmerman site, 50, 67, 78, 83